Springer Theses

Recognizing Outstanding Ph.D. Research

Aims and Scope

The series "Springer Theses" brings together a selection of the very best Ph.D. theses from around the world and across the physical sciences. Nominated and endorsed by two recognized specialists, each published volume has been selected for its scientific excellence and the high impact of its contents for the pertinent field of research. For greater accessibility to non-specialists, the published versions include an extended introduction, as well as a foreword by the student's supervisor explaining the special relevance of the work for the field. As a whole, the series will provide a valuable resource both for newcomers to the research fields described, and for other scientists seeking detailed background information on special questions. Finally, it provides an accredited documentation of the valuable contributions made by today's younger generation of scientists.

Theses are accepted into the series by invited nomination only and must fulfill all of the following criteria

- They must be written in good English.
- The topic should fall within the confines of Chemistry, Physics, Earth Sciences, Engineering and related interdisciplinary fields such as Materials, Nanoscience, Chemical Engineering, Complex Systems and Biophysics.
- The work reported in the thesis must represent a significant scientific advance.
- If the thesis includes previously published material, permission to reproduce this must be gained from the respective copyright holder.
- They must have been examined and passed during the 12 months prior to nomination.
- Each thesis should include a foreword by the supervisor outlining the significance of its content.
- The theses should have a clearly defined structure including an introduction accessible to scientists not expert in that particular field.

More information about this series at http://www.springer.com/series/8790

Xiwang Dong

Formation and Containment Control for High-order Linear Swarm Systems

Doctoral Thesis accepted by
Tsinghua University, Beijing, China

Springer

Author
Dr. Xiwang Dong
School of Automation Science
and Electronic Engineering
Beihang University
Beijing
China

Supervisor
Prof. Yisheng Zhong
Department of Automation
Tsinghua University
Beijing
China

ISSN 2190-5053 ISSN 2190-5061 (electronic)
Springer Theses
ISBN 978-3-662-47835-6 ISBN 978-3-662-47836-3 (eBook)
DOI 10.1007/978-3-662-47836-3

Library of Congress Control Number: 2015944457

Springer Heidelberg New York Dordrecht London

Printed on acid-free paper

Springer-Verlag GmbH Berlin Heidelberg is part of Springer Science+Business Media
(www.springer.com)

A list of articles which have been published before are put in the following:

[1] Xiwang Dong, Jianxiang Xi, Geng Lu, Yisheng Zhong, Formation control for high-order linear time-invariant multi-agent systems with time delays, IEEE Transactions on Control of Network Systems, 2014, 1(3): 232–240. (Reproduced with permission)
[2] Xiwang Dong, Bocheng Yu, Zongying Shi, Yisheng Zhong, Time-varying formation control for unmanned aerial vehicles: Theories and applications, IEEE Transactions on Control Systems Technology, 2015, 23(1): 340–348. (Reproduced with permission)
[3] Xiwang Dong, Fanlin Meng, Zongying Shi, Geng Lu, Yisheng Zhong, Output containment control for swarm systems with general linear dynamics: A dynamic output feedback approach, Systems & Control Letters, 2014, 71(1): 31–37. (Reproduced with permission)
[4] Xiwang Dong, Jianxiang Xi, Geng Lu, Yisheng Zhong, Containment analysis and design for high-order linear time-invariant singular swarm systems with time delays, International Journal of Robust and Nonlinear Control, 2014, 24 (7): 1189–1204. (Reproduced with permission)
[5] Xiwang Dong, Zongying Shi, Geng Lu, Yisheng Zhong, Output containment analysis and design for high-order linear time-invariant swarm systems, International Journal of Robust and Nonlinear Control, 2015, 25(6): 900–913. (Reproduced with permission)
[6] Xiwang Dong, Zongying Shi, Geng Lu, Yisheng Zhong, Time-varying output formation control for high-order linear time-invariant swarm systems, Information Sciences, 2015, 298(20): 36–52. (Reproduced with permission)
[7] Xiwang Dong, Zongying Shi, Geng Lu, Yisheng Zhong, Formation-containment analysis and design for high-order linear time-invariant swarm systems, International Journal of Robust and Nonlinear Control, in press, 2014. (Reproduced with permission)
[8] Xiwang Dong, Jianxiang Xi, Zongying Shi, Yisheng Zhong, Practical consensus for high-order linear time-invariant swarm systems with interaction uncertainties, time-varying delays and external disturbances, International Journal of Systems Science, 2013, 44(10): 1843–1856. (Reproduced with permission)

Supervisor's Foreword

In the past decades, researches on swarm systems consisting of multiple subsystems (also called agents) with interactions have attracted much attention in various fields. In engineering, how to design distributed algorithms that rely on only local interactions for a group of agents working together to achieve certain global group behaviors has become an attractive focus in many scientific communities, especially the control and robotics communities. Compared with a single complex agent, multiple simple agents working cooperatively can obtain great benefits which include low cost, high scalability and flexibility, great robustness, and easy maintenance. This gives rise to a very active and exciting research field cooperative control of swarm systems (or multi-agent systems) which has potential applications in many areas such as cooperative control of intelligent transportation systems, distributed control of power systems, cooperation of multiple robots, distributed optimization of networked systems, formation flying of multiple satellites and unmanned aerial vehicles (UAVs), etc.

In this doctoral thesis, Xiwang Dong investigates formation and containment control problems of high-order linear time-invariant swarm systems with consensus control problems, formation control problems, containment control problems, and formation-containment control problems, which are typical cooperative control problems, discussed respectively. Sufficient conditions for swarm systems with time-varying delays, interaction uncertainties, and external disturbances to achieve practical consensus are presented. Necessary and sufficient conditions for swarm systems to achieve time-varying state/output formations and those for state/output formation feasibilities are proposed respectively. Approaches to specify the motion modes of state/output formation references, to expand feasible state/output formation sets, and to design the state/output formation protocols are proposed respectively. Necessary and sufficient conditions for swarm systems to achieve state/output containment and approaches to design the state/output containment protocols are presented. Sufficient conditions for swarm systems to achieve state/output formation-containment and approaches to design the protocols are shown respectively. State/output consensus problems, state/output consensus tracking problems, state/output formation problems, state/output containment problems, and state formation-containment problems can be regarded as special cases of output formation-containment problems. Theoretical results on formation control are applied to solve the time-varying formation control problems of UAV swarm systems. Autonomous time-varying formation flight experiments using five quadrotor UAVs are performed in outdoor environment.

The above-mentioned results obtained by Xiwang have been appreciated by peer reviewers and editors. For example, the associate editor and reviewers of IEEE Transactions on Control of Network Systems have given the comments that "As all reviewers agree, the paper contains fairly original results" and "The work presented is fairly relevant for control engineering applications as its generality allows to apply it to a considerable pool of scenarios." The reviewers of IEEE Transactions on Control Systems Technology have concluded that the time-varying formation experiments are "remarkable" and "meaningful (recall that most existing work focused on theoretical analysis)." This thesis not only contributes to the theoretical development of cooperative control of swarm systems but also provides practical approaches for the formation and containment control of many swarm systems in engineering.

Beijing
May 2015

Prof. Yisheng Zhong

Abstract

Formation and containment control of swarm systems has broad applications in many fields, such as formation control of unmanned aerial vehicle (UAV) swarm systems, formation control of multiple satellites, and cooperation of multiple robots. This thesis investigates formation and containment control problems of high-order linear time-invariant swarm systems with consensus control problems, formation control problems, containment control problems, and formation-containment control problems discussed respectively. Moreover, theoretical results for formation control of high-order swarm systems are applied to solve the time-varying formation control problems of UAV swarm systems. The main contributions of the thesis are as follow.

(1) The concept of practical consensus is proposed. Sufficient conditions for swarm systems with non-uniform time-varying delays, interaction uncertainties, and time-varying external disturbances which can belong to L_2 or L_∞ to achieve state practical consensus are presented. Explicit expressions of the state practical consensus function and the consensus error bound are also given.
(2) Time-varying state formation control problems for high-order swarm systems with time delays and time-varying output formation control problems for high-order swarm systems are studied respectively. Necessary and sufficient conditions for swarm systems to achieve time-varying state/output formations, necessary and sufficient conditions for state/output formation feasibilities, and explicit expressions of state/output formation reference functions are presented. Approaches to specify the motion modes of state/output formation references, approaches to expand feasible state/output formation sets, and approaches to design the state/output formation protocols are proposed respectively. Necessary and sufficient conditions for UAV swarm systems to achieve time-varying formations and approaches to design the protocol are proposed. Autonomous time-varying formation experiments using five quadrotor UAVs are performed in outdoor environment to demonstrate the theoretical results.

(3) An output containment protocol is proposed based on the dynamic output feedback for high-order swarm systems. Necessary and sufficient conditions for swarm systems to achieve output containment and approaches to design the output containment protocol are presented. For high-order singular swarm systems with time delays, sufficient conditions to achieve state containment are proposed and an approach with less computational complexity to design the protocol is given. It is shown that containment results for singular swarm systems can be applied to solve the consensus tracking problems for singular swarm systems, state containment, and consensus tracking problems for normal swarm systems.
(4) For formation-containment problems of high-order swarm systems, sufficient conditions to achieve state/output formation-containment and approaches to design the protocols are proposed respectively. Necessary and sufficient conditions for high-order swarm systems to achieve state containment are presented as special cases. It is pointed out that state/output consensus problems, state/output consensus tracking problems, state/output formation problems, state/output containment problems, and state formation-containment problems can be regarded as special cases of output formation-containment problems.

Keywords Swarm system · Consensus control · Formation control · Containment control · Formation-containment control

Acknowledgments

First of all, I would like to express my deepest gratitude to my supervisor, Professor Yisheng Zhong, for his guidance, patience, and support in my five-year graduate study at Tsinghua University. His rigorous and conscientious attitude to academic research, diligent and dependable work ethic, easygoing and optimistic attitude to life influenced and benefitted me a lot. Professor Zhong is my great mentor in both academic research and daily life. His edification and exemplary behavior are my lifetime wealth.

Thanks to Professor Zongying Shi and Dr. Geng Lu in the research group for their guidance and help in my study, research, and life at Tsinghua University. Both of them have given lots of constructive suggestions on solving technical problems, revisions of papers, job hunting, and other aspects of life.

I would also like to thank all the previous and current members in the research group of Professor Yisheng Zhong. In the past five years, Dr. Jianxiang Xi, Dr. Ning Cai, Dr. Fanlin Meng, Dr. Lianhua Zhang, Yu Cheng, Furong Lei, Dr. Shusheng Yang, Dr. Xiafu Wang, Dr. Hao Liu, Dr. Yao Yu, Dr. Zhuang Jiao, Bocheng Yu, Yan Zhou and Danjun Li, etc., have provided various forms of care and help. I will never forget the wonderful time we studied and played together.

Last but not the least, I would like to thank my family; that is, my father Qixing Dong, my mother Guimei Liu, my wife Dan Li, and my son Yuce Dong. Without their immeasurable support, constant encouragement, and unwavering love, I would not have been able to finish my thesis.

Publication of this work was financially supported by the National Natural Science Foundation of China (Nos. 61374034, 61333011, 61203071, 61174067 and 60736024) and the Fundamental Research Funds for the Central Universities (No. YWF-14-RSC-101).

Contents

Acronyms

AUV	Autonomous underwater vehilce
BIBS	Bounded input and bounded state
CCL	Core complementarity linearization
CEP	Circular error probable
DoF	Degree-of-freedom
FCS	Flight control system
GCS	Ground control station
GPS	Global positioning system
ILMI	Interactive linear matrix inequality
LMI	Linear matrix inequality
LTI	Linear time-invariant
PBH	Popov-Belevitch-Hautus
PD	Proportional-derivative
QMI	Quadratic matrix inequality
UAV	Unmanned aerial vehicle

Notations

G	Graph
L	Laplacian matrix
$\mathbb{R}^n$	Set of $n \times 1$ real vectors
$\mathbb{R}^{Nn}$	Set of $Nn \times 1$ real vectors
$\mathbb{R}^{N \times n}$	Set of $N \times n$ real matrices
$\mathbb{C}^{N \times n}$	Set of $N \times n$ complex matrices
I	Identity matrix with appropriate dimension
$\mathbf{1}$	Column vector of all ones
e_i	Column vector with 1 as its ith component and 0 elsewhere
0	0 number, 0 vector or 0 matrix with appropriate dimension
$\otimes$	Kronecker product
$\oplus$	Direct sum
$\equiv$	Identically equal
$\Rightarrow$	Implies
$\|x\|$	2-norm of a real vector x
$\mathrm{rank}(A)$	Rank of matrix A
Λ_A	Block diagonal matrix with diagonal blocks A
$\mathrm{Re}(\lambda)$	Real part of number λ
$\mathrm{Im}(\lambda)$	Imaginary part of number λ
Ψ_λ	Matrix $\Psi_\lambda = \begin{bmatrix} \mathrm{Re}(\lambda)I & -\mathrm{Im}(\lambda)I \\ \mathrm{Im}(\lambda)I & \mathrm{Re}(\lambda)I \end{bmatrix}$
A^H	Hermitian transpose of matrix A
A^T	Transpose of matrix A
$\mathscr{L}$	Subscript set of leaders
$\mathscr{F}$	Subscript set of followers
$*$	Symmetry term in symmetric block matrix
Θ	A pseudorandom value with a uniform distribution on the interval $(0, 1)$

Chapter 1
Introduction

Abstract This chapter shows an overview of the whole thesis. First, the scientific and engineering background of formation and containment control is introduced. It is revealed that the theoretical results about formation and containment control have potential applications in formation flying of multiple unmanned aerial vehicles, formation control of multiple satellites, cooperation of multiple robots, etc. Then previous research on consensus control, formation control, containment control, and formation-containment control of swarm systems is reviewed, respectively. Third, the opportunities and challenging in formation and containment control of high-order linear time-invariant swarm systems are given. The research objective of this thesis is proposed. Finally, the main contents and organization of this thesis are presented.

1.1 Scientific and Engineering Background

In the natural world, a biological swarm is formed if multiple biological individuals gather together, such as beast herds, bird flocks, fish shoals, insect swarms, bacteria colonies, etc. It has been discovered by biologists that swarms have more advantages than individuals in finding foods, avoiding predators, migrating, and accomplishing other collective tasks [1–3]. It is well known that the sensing ability, executing ability, and intelligent level of each individual in a swarm are quite limited, and each individual can only accomplish some simple behavior based on the local information it can obtain. However, after forming a large scale of swarm via local interactions and individual movements, the disadvantages of individuals can be overcome and more complicated and powerful macroscopic behavior emerges. This kind of macroscopic behavior is often named as swarming behavior. Figure 1.1 shows three typical swarming behaviors in the natural world; that is, foraging of ant swarm, migrating of bird flock and cruising of fish shoal.

X. Dong, *Formation and Containment Control for High-order Linear Swarm Systems*, Springer Theses, DOI 10.1007/978-3-662-47836-3_1

Fig. 1.1 Typical swarming behaviors in the natural world

Compared with individuals, swarms in the natural world have the following advantages:
(i) By working in groups, the foraging efficiency of a swarm can be improved. For example, an ant swarm can move the food, the weight of which is hundreds of times than the one of a single ant. By attacking in groups, a piranha swarm can eat a muscular bull in a short time.
(ii) By cooperating with each other, individuals in a swarm can gain a much higher survival rate. For example, during the migration of gnu herds, the muscular male gnus walk outside the herd while the female gnus and the young gnus walk inside. In the face of danger, the muscular male gnus lead the whole herd to rush forward like a flash flood, which helps the whole gnu herd out of danger. Marine animals in the deep ocean often locate and hunt their prey by sonar. Millions of small fishes in shoaling, will make their predators mistake them for a huge animal and fear to attack.
(iii) By moving in a swarm, the motion drag can be reduced and the energy for an individual can be saved. Take the migration of greylag geese, for example, by flying in formation, each greylag goose moves in an upwash field that is generated by the wings of the other greylag goose in the formation, which reduces their energy expenditure and increases the migratory distance.

The development of bionics shows that the research on the swarming behavior in the natural world can help us to understand the complicated swarming phenomenons in human society and solve the technical problems in practical engineering. There exist various swarming behaviors in human society, such as the panic of crowds, propagation of fashion, fluctuation of stock market, etc. By studying the swarming behaviors in the natural world, not only the mysteries of the nature can be revealed but also the activities of human beings and the principles of economics can be explained and understood. Moreover, achievements in the research of swarming behavior can provide new concepts, principles, and theories for the development of human technologies. Motivated by the facts stated above, the modeling, analysis, and control of swarming behavior have become important research topics in science and engineering communities in the recent years. Researches define the systems having similar properties as the swarming behavior in the natural world as swarm systems or multiagent systems. In engineering, the agents in a swarm system can be satellites, unmanned aerial vehicles (UAVs), missiles, robots, vehicles, ships, autonomous underwater vehicles (AUVs), routers, sensors, processors, etc., which are control objects with certain coordination ability. In the cooperative control of swarm systems, agents interact locally with each other to achieve the desired macroscopic objective of the swarm system. Up to now, there are many different research branches in the cooperative control of swarm systems which include consensus control [4], formation control [5], containment control [6], formation-containment control [7], consensus tracking control [8], flocking control [9], pursuit-evasion control [10], distributed filtering [11], etc. Formation and containment control in this thesis consists of the first four research branches. Consensus control is the foundation of formation control and also a special case of formation control. Formation control and containment control are foundation of formation-containment control. Moreover, consensus control, formation control, and containment control can all be regarded as special cases of formation-containment control. The results on formation and containment control have broad potential applications in various areas, such as formation control of satellites [12], UAVs [13], and missiles [14], and cooperation of multiple robots [15], wireless sensor networks [16], etc. Several application examples are given as follows:

(1) Formation control of UAV swarm systems
UAV is a kind of reusable unmanned aerial vehicle that can carry various instruments and equipments. The typical features of the UAV include small size, light weight, high maneuverability, low cost, easy to hide, and zero casualty. With the expansion of the applications of UAVs, the requirement for the carrying capacity and endurance capacity is increasing. The drawbacks of single UAV in accomplishing complicated tasks are emerging. It is becoming an inevitable trend to employ multiple UAVs to perform tasks by cooperation. There are many advantages for a UAV swarm system to flight in formation. For example, for multiple fixed wing UAVs to fly in specific formation, the drag and fuel consumption can be reduced significantly. As a result, the flight range or the payload of the UAV swarm system can be increased and some tasks that are impossible for single UAV can be achieved. An autonomous formation flight project has been implemented at the NASA Dryden flight research center, to demonstrate the benefits of formation flight using two F/A-18 aircrafts. It was shown that the drag can be reduced by more than 20 % and more than 18 % fuel can be saved by flying in formation [17]. In 2012, Airbus proposed the concept of smarter skies and pointed out that multiple aircrafts can fly in formation to reduce drag and lower energy use from 2050 [18]. In the military area, multiple UAVs with electronic countermeasure equipments can fly in certain formation to intercept, jam, and deceive the radio signal of the ground radar station [19, 20]. Multiple UAVs with telecommunication relay devices can form a temporary wireless network above a region that requires wide communication bandwidth to transmit texts, images, audio, and video information [21]. Figure 1.2 shows an example of formation flight of a quadrotor swarm system at Tsinghua University.

Fig. 1.2 Quadrotor swarm systems fly in formation at Tsinghua University

(2) Formation control of multiple satellites

In 1990s, some researchers proposed the concept of virtual satellite and pointed out that the function of a huge satellite can be realized by having multiple much simpler and smaller satellites working in formation [22]. Compared with a single complex and big satellite, the virtual satellite has the following advantages. First, the miniature satellites in the virtual satellite can be distributed in a broad space, which can provide large aperture and baseline for measurement. By dealing with the information of all small satellites at different locations using data fusion techniques, the measurement accuracy of the virtual satellite can be improved significantly, which is helpful for deep space exploration, navigation and localization, and three-dimensional imaging. Second, since the function of the virtual satellite is realized by many miniature satellites, in the case where certain miniature satellite got damaged or destroyed, the virtual satellite can still work in a degraded mode by reconfiguring the coordination relationship among the miniature satellites rather than collapse. Moreover, because miniature satellites are much easier to launch, new satellites can be replenished in a short time, which greatly enhances the robustness and maintainability of the whole system. Finally, miniature satellites have short product development cycle, and can be produced and launched in a batch way, which reduces the launch difficulty, risk, and cost significantly. Figure 1.3 shows the concept of formation-flying satellites proposed by DARPA [23].

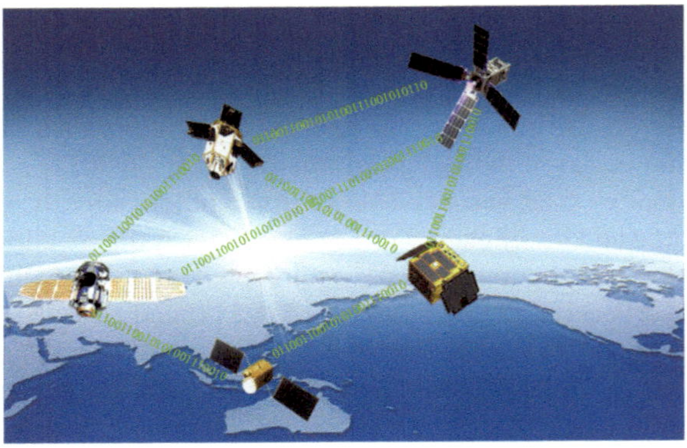

Fig. 1.3 Concept of formation-flying satellites proposed by DARPA

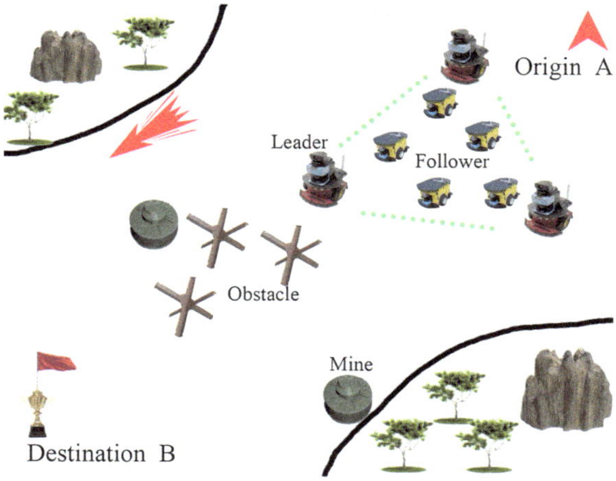

Fig. 1.4 Cooperative migration of multiple robots in the hazardous environment

(3) Corporation of multiple robots

The robot is a kind of new production tool that can take the place of humans in dangerous environments and repetitive manufacturing processes. The invention of robots enhances the ability of humans to transform the world and accelerate the development of the society. Nowadays, robots can be found from distant sky to abysmal sea, and also from high-tech laboratory to the average family. Although the intelligence and capability of single robot are improving, with the expansion of the application areas, it is becoming a new trend for multiple robots to work together to accomplish certain complex tasks. For example, in airports, harbors, and marshaling yards, large tonnage or jumbo size objects need multiple robots working together to carry [24]. In robot soccer games, complex actions including double pass, pass and receive combinations, and man-for-man marking can only be accomplished by the corporation of multiple robots. Figure 1.4 depicts a sketch that multiple robots migrate in a hazardous environment using a cooperative method. In Fig. 1.4, there are many obstacles and mines between the origin and the destination. To make sure that the whole robot swarm system moves safely to the destination, one simple strategy is to equip all the robots with necessary sensors to detect the obstacles and mines. However, if the number of robots is very large, this strategy will cost a lot. On the other hand, only part of the robots need to be equipped with the sensors, and define those as leaders while the ones without sensors as followers. Using containment control approaches, it can be guaranteed that all the followers stay in the safety area formed by the leaders. Therefore, the whole robot swarm system can get to the destination safely with a much less cost [15].

1.2 Literature Review on Formation and Containment Control

In the research field of cooperative control for swarm systems, consensus control, formation control, containment control, and formation-containment control are closely related with each other. It is a gradual process for the research moves from consensus control to formation control, and also from formation control and containment control to formation-containment control. Although many research topics in cooperative control of swarm systems which include consensus control, formation control, and containment control can be unified in the framework of formation-containment control, the research on formation-containment control is based on the results obtained in the consensus control, formation control, and containment control. The formation and containment control problems discussed in this thesis consist of consensus control problem, formation control problem, containment control problem, and formation-containment control problem. In the following, the research status on these four problems will be reviewed, respectively. It should be pointed out that the theory on formation control is also applied to formation control of UAVs and experimental results are also presented in this thesis. Therefore, the research status of formation control for UAV swarm systems is also summarized.

1.2.1 Consensus Control of Swarm Systems

Consensus control problem is one of the most fundamental and important problems in cooperative control of swarm systems. A swarm system is said to achieve consensus if all the agents reach an agreement on certain variables of interest, where the variables can be named as coordination variables. To achieve consensus, agents in a swarm system often interact with each other locally. For each agent, the local interaction is realized by constructing distributed controller or protocol using neighboring relative information.

Although consensus control problem has already been a hot research topic, it is not a new problem. In 1980s, Borkar and Varaiya [25], and Tsitsiklis and Athans [26] from the computer science community studied the asynchronous consensus problems in the distributed computation. In the year of 1995, Vicsek et al. [27] proposed a discrete-time model (Vicsek model) for self-propelled particle swarms in which all agents move in the same plane with the same speed but different headings. It is shown that the particle swarm can reach an agreement on heading via a local update rule. Based on graph theory and matrix theory, Jadbabaie et al. [28] studied the consensus behavior of the Vicsek model theoretically, and proved that if the interaction topology is jointly connected then the heading of the swarm system will achieve consensus.

Olfati-Saber and Murray [4] investigated consensus problems for first-order continuous-time swarm systems. For a *first-order swarm system* with N agents, the dynamics of each agent can be described by the following single-integrator model

$$\dot{x}_i(t) = u_i(t) \ (i = 1, 2, \ldots, N), \tag{1.1}$$

where $x_i(t)$ is the state of the ith agent (coordination variable) and $u_i(t)$ is the control input (protocol) of the ith agent. If $\lim_{t\to\infty}(x_i(t) - x_j(t)) = 0 \ (i, j = 1, 2, \ldots, N)$, then swarm system (1.1) is said to achieve consensus. In the case where the interaction topology is fixed, the following protocol is proposed in [4]

$$u_i(t) = \sum_{j\in N_i} w_{ij} \left(x_j(t) - x_i(t)\right), \tag{1.2}$$

where N_i represents the neighbor set of the ith agent and w_{ij} stands for the interaction strength from agent j to agent i. Olfati-Saber and Murray [4] proved that if the interaction topology is strongly connected and balanced, then swarm system (1.1) with protocol (1.2) achieves consensus. Based on the work of [4] and [28], Ren and Beard [29] discussed consensus problems for first-order swarm systems with directed interaction topologies, and showed that if the interaction topology contains a spanning tree, then consensus can be achieved. Using H_∞ control approach, Lin et al. [30] addressed consensus problems for first-order swarm systems with interaction uncertainties and external disturbances, and presented sufficient conditions for swarm systems to achieve consensus. In [31], constant and time-varying communication delays were considered for first-order swarm systems with undirected interaction topologies, and necessary and sufficient conditions were derived. Tian and Liu [32] studied consensus problems for first-order swarm systems with communication delays and input delays. Criteria for first-order swarm systems with time-varying delays and switching topologies to achieve consensus were proposed in [33–35]. Moreover, consensus problems for first-order swarm systems with random switching interconnection topologies [36, 37], finite-time consensus problems [38, 39], and asynchronous consensus problems [40–42] for first-order swarm systems have also been studied.

In the aforementioned research, the dynamics of each agent is restricted to be first order. In recent years, consensus problems for swarm systems have been studied extensively, and lots of results have been obtained. In the following, the research on consensus for second-order swarm systems and high-order swarm systems will be reviewed, respectively.

(1) Consensus control for second-order swarm systems

In some practical application, the dynamics of each agent can be described by second-order model, such as the multivehicle systems and multirobot systems [43]. Therefore, it is meaningful to study consensus problems for second-order swarm systems.

For a *second-order swarm system*, the dynamics of agent i can be described by

$$\begin{cases} \dot{x}_i(t) = v_i(t), \\ \dot{v}_i(t) = u_i(t), \end{cases} \tag{1.3}$$

where $x_i(t)$ and $v_i(t)$ are the position and velocity, and $u_i(t)$ is the control input. Xie and Wang [44] constructed the following consensus protocol:

$$u_i(t) = k_v v_i(t) + \sum_{j \in N_i} w_{ij}(x_j(t) - x_i(t)), \tag{1.4}$$

where k_v is the constant gain. They proved that if the interaction topology is connected and $k_v < 0$, then second-order swarm system (1.3) with protocol (1.4) reaches agreement on the position and velocity. Ren and Atkins [43] proposed a protocol as

$$u_i(t) = \sum_{j \in N_i} w_{ij} \left(x_j(t) - x_i(t) + \gamma_v \left(v_j(t) - v_i(t) \right) \right), \tag{1.5}$$

where $\gamma_v > 0$ is a uniformly bounded constant gain. It can be found in [43] that if the interaction topology contains a spanning tree and γ_v satisfies

$$\gamma_v > \max_{\lambda_i \neq 0} \sqrt{\frac{2}{|\lambda_i| \cos\left(\frac{\pi}{2} - \tan^{-1} \frac{-\mathrm{Re}(\lambda_i)}{\mathrm{Im}(\lambda_i)}\right)}},$$

then swarm system (1.3) with protocol (1.5) achieves consensus, where λ_i $(i = 2, 3, \ldots, N)$ are the nonzero eigenvalues of $-L$, L is the Laplacian matrix, $\mathrm{Re}(\lambda_i)$ and $\mathrm{Im}(\lambda_i)$ stand for the real part and imaginary part of λ_i. Consensus problems for second-order swarm systems with switching topologies were discussed in [45–48]. In [49–52], time delays were considered. Robust consensus problems for second-order swarm systems with external disturbances and parameter uncertainties were investigated in [53, 54]. Consensus control for data-sampled second-order swarm systems [55–57] and second-order swarm systems with nonlinear dynamics [58–60] are also interesting topics.

(2) Consensus control for high-order swarm systems
In the first-order and second-order swarm systems, the dynamics of each agent has special structure; that is, single integrator and double integrators. Therefore, the analysis and design for first-order and second-order swarm systems are much easier. In the past decade, numerous results on consensus control of first-order and second-order swarm systems have been obtained. However, in practical applications, the dynamics of some swarm systems can only be described by models with high order, such as the UAV swarm systems with six degrees of freedom (DoF) and robotic manipulator swarm systems with multiple joints. Moreover, first-order and second-order swarm systems can be regarded as special cases of high-order swarm systems.

Therefore, it is more practical and general to study consensus problems for high-order swarm systems.

A *high-order linear time-invariant (LTI) swarm system* can be described by

$$\begin{cases} \dot{x}_i(t) = Ax_i(t) + Bu_i(t), \\ y_i(t) = Cx_i(t), \end{cases} \tag{1.6}$$

where $i = 1, 2, \ldots, N$, $A \in \mathbb{R}^{n\times n}$, $B \in \mathbb{R}^{n\times m}$, $C \in \mathbb{R}^{q\times n}$, and $x_i(t) \in \mathbb{R}^n$, $y_i(t) \in \mathbb{R}^q$, $u_i(t) \in \mathbb{R}^m$ are the states, outputs, and control inputs, respectively. If for any $i, j \in \{1, 2, \ldots, N\}$, $\lim_{t\to\infty}(x_i(t) - x_j(t)) = 0$ or $\lim_{t\to\infty}(y_i(t) - y_j(t)) = 0$, then swarm system (1.6) is said to achieve state consensus or output consensus. Xiao and Wang [61] proposed the following protocol:

$$u_i(t) = K_1 x_i(t) + K_2 \sum_{j\in N_i} w_{ij} \left(x_j(t) - x_i(t)\right), \tag{1.7}$$

where K_1 and K_2 are gain matrices with appropriate dimension, and presented necessary and sufficient conditions for high-order swarm system (1.6) with protocol (1.7) to achieve consensus. For swarm system (1.6) with undirected interaction topologies, approaches to determine the gain matrices K_1 and K_2 were given in [62]. Xi et al. [63] proposed a state space decomposition approach to deal with the consensus problems for high-order swarm systems, and presented necessary and sufficient conditions and explicit expressions for consensus function. In [64], the following consensus protocol was constructed based on static output feedback:

$$u_i(t) = K \sum_{j\in N_i} w_{ij} \left(y_j(t) - y_i(t)\right), \tag{1.8}$$

where K is the gain matrix, and necessary and sufficient conditions for swarm system (1.6) to achieve state consensus were proposed. Li et al. [65] proposed a consensus protocol based on dynamic output feedback, and presented criteria for swarm system (1.6) to achieve state consensus based on separation principle and consensus region approaches. Based on [65], Li et al. [66] designed an distributed consensus protocol using adaptive approaches which does not require the global information of the interaction topology. Seo et al. [67] studied state consensus problems for high-order swarm systems using low-gain approaches. Time delays were considered in [68, 69]. The effects of switching topologies on the consensus of high-order swarm systems were addressed in [70, 71]. Robust consensus problems for high-order swarm systems with external disturbances were discussed in [72, 73]. Other research topics include output consensus for high-order swarm systems [74, 75], state consensus for singular high-order swarm systems [76, 77], consensus for heterogeneous high-order swarm systems [78, 79], etc.

1.2.2 Formation Control of Swarm Systems

Formation control is not a new topic and has been studied a lot in traditional robotics community. Many centralized and decentralized formation control approaches have been proposed in the past decades [80]. According to the fundamental ideas in control schemes, traditional formation control approaches can be roughly classified into leader–follower-based formation control strategy [81], behavior-based formation control strategy [82], and virtual structure-based formation control strategy [83]. In the leader–follower-based formation control, one or multiple agents are specified as leaders and the rest are followers. Leaders follow the desired trajectory, and followers keep certain distance and angle offsets with respect to leaders. The basic idea behind the behavior-based formation control is to prescribe several desired behaviors for each agent, such as formation keeping, collision avoidance, obstacle avoidance, and goal seeking, and to make the control action of each agent a weighted average of the control for each behavior. By adjusting the wights for different behaviors, the desired behavior can be achieved by the swarm system. In the virtual structure-based formation control, the desired formation is treated as a virtual rigid structure and each agent is treated as a node in the virtual structure. Each agent tracks the movement of the corresponding node in the virtual structure to achieve the formation. These classic formation control approaches have their advantages and disadvantages [84]. The advantage of leader–follower-based approach is that it is simple and easy for implementation. The disadvantage, however, is that it lack robustness, and the failure of the leader may destroy the whole formation. Moreover, there is no explicit feedback to the formation, which means that if the leader moves too fast, it may be difficult for the followers to track. The strength of the behavior-based approach is that it can achieve multiple control objectives simultaneously and it has explicit feedback to the formation. The primary weakness is that group behavior cannot be explicitly defined, and it is difficult to analyze mathematically. The strength of the virtual structure-based approach is that it has feedback to the virtual structure and can achieve the precise formation. The disadvantage is that it consumes too many computations and communications to synthesize the virtual structure and track the waypoint.

With the development of cooperative control of swarm systems, especially the development of consensus theory, more and more researchers realize that consensus theories can be applied to deal with the formation control problems of swarm systems. As a new formation control approach, consensus-based formation control strategy is attracting more and more attention from robotics and control communities. The basic idea behind the consensus-based formation control is that the states or outputs of all agents keep certain offsets corresponding to a common formation reference which is previously unknown to each agent. By local interaction among neighboring agents, the states or outputs of all agents with respect to the formation vectors reach an agreement on the formation reference, which means the formation is achieved by the swarm system. One of the most used approaches in dealing with consensus-based formation control problems is to construct appropriate coordinate transformation and convert the formation problem into the consensus problem which can be solved by the consensus theory. Ren [85] applied the consensus protocol (1.5) to study the formation control problem of second-order swarm systems, and proved that leader–follower, behavior and virtual structure-based formation control approaches can be regarded as special cases of consensus-based formation control approaches. Besides, the weakness of these approaches can be overcome. In [86], experiments were carried out on multirobot systems to show the effectiveness of consensus-based formation control approaches. Xiao et al. [39] proposed a finite-time formation control protocol for first-order swarm systems, and proved that the constant formation can be achieved in a finite time. Xie and Wang [87] investigated formation control problems for second-order swarm systems with undirected interaction topologies, and presented sufficient criteria to achieve constant formations. The effects of time delays on the formation control of second-order swarm systems were addressed in [88].

In [39, 85–88], the dynamics of each agent is restricted to be first order or second order. Using consensus-based formation control strategy, Lafferriere et al. [89] studied the time-invariant formation control problems for a special high-order swarm systems which can be regarded as a series of second-order models. It should be pointed out that this special model can simplify the the formation control problems significantly. Fax and Murry [5] discussed the formation stability problems for general high-order LTI swarm systems described by (1.6). Porfiri et al. extended the results of [5] by introducing a measurable vector field for swarm systems, and dealt with formation and tracking control problems for general high-order LTI swarm systems simultaneously. Although the dynamics of each agent in [5] and [90] is generally high-order, only formation stability problems were addressed and the problem that how to achieve desired formations was not considered. For a given swarm system, whether a desired formation is feasible is a essential problem. Lin et al. [91] investigated the formation feasibility problems for multirobot systems. Ma and Zhang [92] proposed a formation control protocol as

$$u_i(t) = K_1 \sum_{j\in N_i} w_{ij} \left(x_j(t) - x_i(t) - (h_j - h_i)\right) \; (i = 1, 2, \ldots, N), \tag{1.9}$$

where K_1 is the gain matrix with appropriate dimension and h_i is a constant vector that describes the desired formation, and studied the formation feasibility problems for high-order LTI swarm systems. It should be pointed out that in [92], the formation is time-invariant, and the feasible formation set is quite limited. Dong et al. [93] presented necessary and sufficient conditions for high-order swarm systems to achieve time-varying state formations, necessary and sufficient conditions for time-varying formation feasibilities, and approaches to expand the feasible time-varying formation set.

The aforementioned work focuses on the theoretical results on formation control, especially consensus-based formation control. With development of formation control theory, how to apply the theoretical results to deal with the formation control problems of practical swarm systems, such as UAV swarm systems, robot swarm systems, AUV swarm systems, etc., becomes a critical issue. Since the theoretical results obtained in this thesis are applied to UAV swarm systems, a literature review on formation control of UAV swarm systems will be summarized in the following.

Formation control of UAV swarm systems will encounter the problems caused by the dynamics property of UAVs and the common problems of general formation control. Typical UAVs can be classified into fixed wing UAVs and rotary wing UAVs according to the generation of lift. Rotary wing UAVs consists of the single rotor UAV and the multirotor UAV. A UAV in three-dimensional space can be treated as a rigid body with six DoFs. The six DoFs include three DoFs in translation and three DoFs in rotation. In the previous research for formation control of UAV swarm systems, the dynamics of each UAV was often described by first-order model, second-order model, three DoFs nonholonomic model, six DoFs nonholonomic model, and linearised high-order model, and the formation control theory was based on leader–follower, behavior, virtual structure, and consensus approaches as stated above. Those five models and four formation control strategies were combined with adaptive control, fuzzy control, sliding mode control, optimal control, and robust control theories to investigate the formation control problems of UAV swarm systems see, e.g., [13, 94–99]. However, in those literature, the formation control theories were demonstrated by numerical simulations.

Experiments on formation control of UAV swarm systems mainly use leader–follower, virtual structure, and consensus-based approaches. Leader–follower-based formation control problems for fixed wing UAV swarm systems with three DoFs nonholonomic dynamics were studied by Gu et al. [100], and formation experiments were performed with one leader UAV and two follower UAVs. Yun et al. [101] discussed formation control problems for single rotor UAV swarm systems with linearised high-order model using leader–follower-based approaches, and validated their results by two single rotor UAVs. You and Shim [102] combined leader–follower-based formation control approaches with proportional-derivative (PD) controllers to investigate the formation control problems of fixed wing UAV swarm systems, and showed formation flight experiments using two fixed wing UAVs. Leader–follower-based formation controller were used to control the formation flight of two fixed wing UAVs with six DoFs nonholonomic model by Di et al. [103]. Kushleyev et al. [104] studied formation control problems for quadrotor UAV swarm systems with six DoFs nonholonomic dynamics based on virtual structure approaches, and performed series of indoor formation experiments. Constant formation experiments on quadrotor UAV swarm systems based on consensus approaches can be found in [105] and [106]. Yu et al. [107] applied consensus-based time-varying formation control theories to the formation control of quadrotor swarm systems, and carried out autonomous time-varying formation experiments using three quadrotor UAVs in outdoor environment.

1.2.3 Containment Control of Swarm Systems

The object of containment control is to drive the states or outputs of followers to converge to the convex hull spanned by the states or outputs of leaders. Containment control is inspired by some swarm behavior in the natural world and the development of consensus theories. On the one hand, biologists have observed that in the mating season of silkworm moths, the females intermittently release the pheromone to attract males, and the males are attracted to a general area formed by the females, which helps them to mate [108, 109]. On the other hand, in the early research stage of consensus control, there were no specify leaders in a swarm system. With the development of research, researchers realized that there could be one or multiple leaders providing global state references or external command inputs for the whole swarm system. Consensus with one leader is often regarded as consensus tracking. Consensus tracking problems for first-order, second-order, and high-order swarm systems have already been studied extensively [110–113]. Under this circumstance, Ji et al. [114] proposed a stop-go policy for first-order swarm systems to achieve containment. Meng et al. [115] discussed finite-time containment control problems for swarm systems with rigid body agents. Notarstefano et al. [116] constructed the following containment protocol for first-order swarm systems described by (1.1),

$$\begin{cases} u_i(t) = 0, i \in Leader, \\ u_i(t) = -\sum\limits_{j \in N_i(t)} (x_i(t) - x_j(t)), i \in Follower, \end{cases} \tag{1.10}$$

where $N_i(t)$ is the switching neighbor set of follower i. They proved that under protocol (1.10), swarm system (1.1) achieves containment if the interaction topology is jointly connected. Containment problems for first-order and second-order swarm systems with both stationary and dynamic leaders were addressed in [15] and [117]. Liu et al. [118] proposed necessary and sufficient conditions for first-order and second-order swarm systems to achieve containment. Lou and Hong [119] investigated containment problems for second-order swarm systems with randomly switching topologies. The dynamics of each agent in [114–119] is restricted to be first order or second order. By classifying agents into the internal agents and boundary agents, Liu et al. [120] presented necessary and sufficient conditions for the states of internal agents to converge to the convex hull spanned by the states of boundary agents. Sufficient conditions for high-order LTI swarm systems to achieve containment were given in [121]. State containment problems for high-order LTI singular swarm systems with time delays and output containment problems for high-order LTI swarm systems were studied in [122] and [123].

1.2.4 Formation-Containment Control of Swarm Systems

Formation-containment control is a new problem related with both formation control and containment control. In [114–123], it is assumed that there exist no interactions among leaders. In some practical applications, leaders not only provide global state references but also achieve certain complicated tasks via local interactions, such as the formation control. A swarm system is said to achieve state or output formation-containment if the states or outputs of leaders achieve predefined formation and at the same time the states or outputs of followers stay in the convex hull spanned by those of leaders. It should be pointed out that consensus problem, consensus tracking problem, formation control problem, and containment control problem can all be regarded as special cases of formation-containment control problems. However, due to the limitation of formation control and containment control, only quite a few research results on formation-containment control can be found. Ferrari-Trecate et al. [124] addressed the concept of formation-containment. Dimarogonas et al. [7] studied the formation-containment problems for nonholonomic robot swarm systems and presented sufficient conditions to achieve formation-containment. In [7, 124], the dynamics of each agent is first order or second order. Formation-containment control problems for high-order LTI swarm systems are still open.

1.3 Opportunities and Challenges

From the literature review in Sect. 1.2, one sees that there are still many opportunities and challenges in consensus control, formation control, containment control, and formation-containment control of high-order LTI swarm systems. These opportunities and challenges are also the problems to be studied by this thesis.

(1) Problems in consensus control of swarm systems
In practical swarm systems, there may exist nonuniform time-varying delays, interaction uncertainties and external disturbances. Under the persistent effect of external disturbances, it may be difficult to achieve the precise consensus. Consensus problems for high-order swarm systems with nonuniform time-varying delays, interaction uncertainties, and external disturbances have not been studied before. Moreover, consensus function is a tool to describe the macroscopic movement of the swarm system. However, explicit expression of the consensus function for high-order swarm systems with nonuniform time-varying delays, interaction uncertainties, and external disturbances has not been obtained in the existing literature.

(2) Problems in formation control of swarm systems
First, in most of the literature, the desired formation is time-invariant. Time-varying state/output formation control problems and state/output formation feasibility problems for high-order swarm systems are still open. For a given high-order swarm system, (i) under what conditions the time-varying state/output formation is feasible; (ii) how to determine the explicit expression of the time-varying state/output formation reference function; (iii) how to assign the motion modes of the state/output formation reference function; (iv) how to expand the feasible state/output formation set; and (v) how to design the time-varying state/output formation protocol are all challenging problems.

Second, time delays are inevitable in practical swarm systems. For high-order swarm systems with time delays, (i) under what conditions the time-varying state formation can be achieved; (ii) what are the effects of time delays on the state formation feasibility, feasible formation set and state formation reference; and (iii) how to design the time-varying state formation protocol under time delays need further study.

Third, how to apply the theoretical results on time-varying formation control of high-order swarm systems to the practical UAV swarm systems, and carry out autonomous time-varying formation experiments using multiple UAVs are meaningful problems.

(3) Problems in containment control of swarm systems
First, necessary and sufficient conditions for first-order and second-order swarm systems to achieve state containment, and sufficient conditions for high-order swarm systems to achieve state containment have been obtained. However, for high-order swarm systems where the motion modes of leaders can be specified, how to obtain the necessary and sufficient conditions to achieve state containment is still a difficult problem.

Second, if each agent can only obtain neighboring output information and only the outputs of followers are required to converge to the convex hull spanned by those of leaders, then output containment control problems arise. Output containment analysis and design problems for high-order swarm systems have not be studied.

Third, singular swarm system is a kind of swarm system where there exist algebraic constraints among states [76, 77]. Singular swarm system is more general than the normal swarm system in describing the practical systems. So far, containment control for high-order singular swarm systems has not been reported. Containment analysis and design problems for high-order singular swarm systems with time delays are still open.

(4) Problems in formation-containment control of swarm systems
In the previous results on formation-containment, the dynamics of each agent is restricted to be first-order or second-order. State/output formation-containment analysis and design problems for high-order swarm systems need to be investigated.

1.4 Contents and the Outline

In this section, the main contents of this thesis are summarized and the outline of this thesis is presented.

1.4.1 Contents

This thesis studies the consensus control problem, formation control problem, containment control problem, formation-containment control problem of high-order LTI swarm systems. The main contents are as follows:

(1) Contents on consensus control
The effects of nonuniform time-varying delays, interaction uncertainties and time-varying external disturbances belonging to L_2 or L_∞ are studied for high-order swarm systems to achieve practical consensus. Using the state space decomposition approach, practical consensus problems of the swarm system are converted into stability problems of the disagreement subsystem. Based on the Lyapunov–Krasovskii

functional approach and the linear matrix inequality technique, sufficient conditions for swarm systems to achieve practical consensus are proposed. An explicit expression of the practical consensus function and consensus error bound under L_2 or L_∞ disturbances are presented.

(2) Contents on formation control
First, time-varying state formation control problems for high-order swarm systems with time delays are studied. Based on state transformation and state space decomposition approaches, necessary and sufficient conditions for high-order swarm systems with time delays to achieve time-varying formations, and necessary and sufficient conditions for time-varying state formation feasibilities are proposed. An explicit expression of the time-varying state formation reference function is derived, and approaches to specify the motion modes of the state formation reference function are given. Approaches to expand the feasible time-varying state formation set and design the formation control protocol are presented.

Second, time-varying output formation control problems for high-order swarm systems are discussed. Based on model transformation and output space decomposition approaches, and partial stability theory, necessary and sufficient conditions for swarm systems to achieve time-varying output formations, and necessary and sufficient conditions for time-varying output formation feasibilities are presented. An explicit expression of the time-varying output formation reference function and approaches to partially assign the motion modes of the output formation reference are proposed. Approaches to expand the feasible time-varying output formation set and determine the gain matrices in the output formation control protocol are given.

Third, theories on time-varying formation control of high-order swarm systems are applied to solve the time-varying formation control problems of UAV swarm systems. Necessary and sufficient conditions for UAV swarm systems to achieve time-varying formations and approaches to design the formation control protocol are proposed. Series of autonomous time-varying formation control experiments are performed using five quadrotor UAVs in outdoor environment.

(3) Contents on containment control
First, for high-order swarm systems where the motion modes of leaders can be assigned, necessary and sufficient conditions for swarm systems to achieve state containment are presented, and approaches to design the containment control protocol are given.

Second, output containment control problems for high-order swarm systems are studied using a dynamic output feedback approach. An output containment protocol is constructed using the outputs of neighboring agents. Necessary and sufficient conditions for high-order swarm systems to achieve output containment are presented. Moreover, approaches to design the dynamic output containment protocol are presented by solving an algebraic Riccati equation.

Third, state containment control problems for high-order singular swarm systems with time delays are investigated. Sufficient conditions for swarm systems to achieve state containment are presented, and approaches to design the time-delayed protocol are proposed. Results on containment control of singular high-order swarm systems

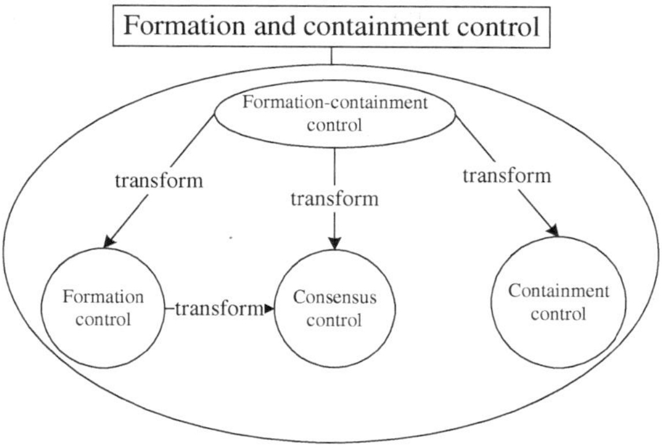

Fig. 1.5 Relationship among formation-containment control, consensus control, formation control and containment control

can also be applied to deal with the consensus tracking problems for singular and normal high-order swarm systems with time delays, and containment problems for normal high-order swarm systems with time delays.

(4) Contents on formation-containment control
Using model transformation approaches, formation-containment control problems for high-order swarm systems are transformed into asymptotic stability problems. Sufficient conditions for high-order swarm systems to achieve state/output formation-containment are proposed. Explicit expressions of the time-varying state/output

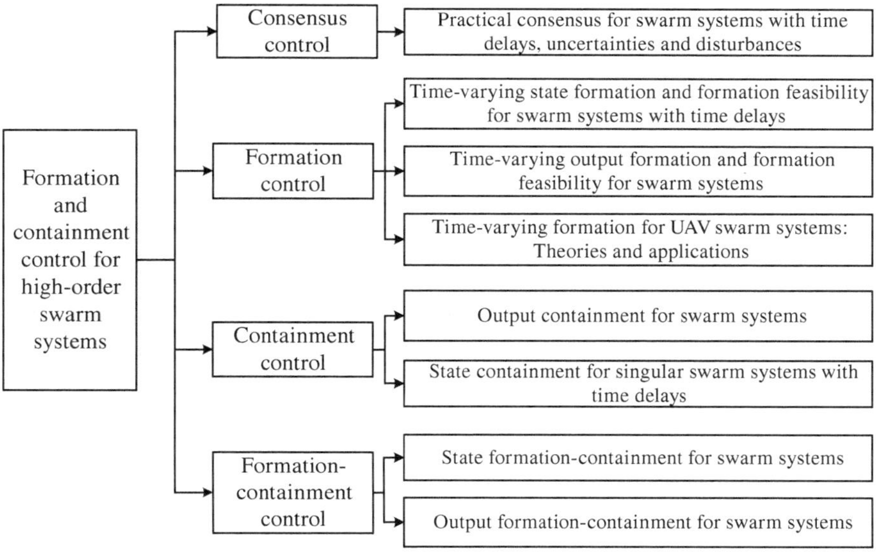

Fig. 1.6 Main problems studied in this thesis

formation reference functions are presented, and approaches to design the state/output formation-containment protocols are given. It is shown that state/output consensus control, state/output consensus tracking control, state/output formation control and state/output containment control can all be regarded as special cases of state/output formation-containment control.

The relationship among formation-containment control, consensus control, formation control and containment control investigated in this thesis is revealed in Fig. 1.5. The main problems studied in this thesis are shown in Fig. 1.6.

1.4.2 Outline of This Thesis

This thesis includes seven chapters.

Chapter 1 **Introduction** The scientific and engineering background of formation and containment control, literature review, the main contents and outline of the thesis are summarized.

Chapter 2 **Preliminaries** Basic definitions and preliminary results on graph theory, consensus decomposition of linear space theory, matrix theory, linear system theory and singular system theory are introduced.

Chapter 3 **Consensus control of swarm systems** The concept of practical consensus is proposed to describe the consensus of swarm systems with external disturbances. Sufficient conditions for swarm systems to achieve practical consensus under the effects of nonuniform time-varying delays, interaction uncertainties and external disturbances are presented, and explicit expressions of the practical consensus function are given.

Chapter 4 **Formation control of swarm systems** For time-varying state formation control problems of high-order swarm systems with time delays and time-varying output formation control problems of high-order swarm systems, necessary and sufficient conditions for swarm systems to achieve time-varying state/output formations are proposed, and necessary and sufficient conditions for state/output formation feasibilities are presented. Explicit expression of the state/output formation reference function, and approaches to specify the motion modes of the state/output formation reference are provided. Approaches to expand the feasible time-varying state/output formation sets and design the state/output formation protocols are given. Theoretical results are applied to the time-varying formation control of UAV swarm systems. Time-varying formation flight experiments are carried out using five quadrotor UAVs.

Chapter 5 **Containment control of swarm systems** Output containment problems for high-order swarm systems are studied using a dynamic output feedback protocol. Necessary and sufficient conditions for swarm systems to achieve output containment and approaches to design the output containment protocol are proposed. State containment problems for high-order singular swarm systems with time delays are investigated. Sufficient conditions for singular swarm systems to achieve state containment and approaches to design the time-delayed state containment protocol are presented.

Chapter 6 **Formation-containment control of swarm systems** State/output formation-containment problems for high-order swarm systems are addressed. Sufficient conditions for swarm systems to achieve state/output formation-containment and approaches to design the protocols are proposed. Necessary and sufficient conditions for swarm systems to achieve state containment and time-varying output formations are derived as special cases. It is pointed out that state/output consensus control, state/output consensus tracking control and state/output formation control can be unified in the framework of state/output containment control.

Chapter 7 **Conclusions and future work** The whole work of this thesis is summarized and some remaining open problems that require further investigation are discussed.

1.5 Conclusions

In this chapter, first, the backgrounds and motivations for formation and containment control of high-order swarm systems were addressed. Then literature reviews on consensus control, formation control, containment control, and formation-containment control of swarm systems were summarized, respectively, and challenging problems to be studied were pointed out. Finally, the main contents and outline of this thesis were introduced.

References

1. Vicsek T (2001) A question of scale. Nature 411(24):421
2. Weimerskirch H, Martin J, Clerquin Y et al (2001) Energy saving in flight formation. Nature 413(18):697–698
3. Couzin ID, Krause J, Franks NR et al (2005) Effective leadership and decision-making in animal groups on the move. Nature 433(3):513–516
4. Olfati-Saber R, Murray RM (2004) Consensus problems in networks of agents with switching topology and time-delays. IEEE Trans Autom Control 49(9):1520–1533
5. Fax JA, Murray RM (2004) Information flow and cooperative control of vehicle formations. IEEE Trans Autom Control 49(9):1465–1476
6. Ji M, Ferrari-Trecate G, Egerstedt M et al (2008) Containment control in mobile networks. IEEE Trans Autom Control 53(8):1972–1975
7. Dimarogonas DV, Egerstedt M, Kyriakopoulos KJ (2006) A leader-based containment control strategy for multiple unicycles. In: Proceedings of the 45th IEEE conference on decision and control, pp 5968–5973
8. Hong YG, Hu JP, Gao LX (2006) Tracking control for multi-agent consensus with an active leader and variable topology. Automatica 42(7):1177–1182
9. Olfati-Saber R (2006) Flocking for multi-agent dynamic systems: algorithms and theory. IEEE Trans Autom Control 51(3):401–420
10. Kim TH, Sugie T (2007) Cooperative control for target-capturing task based on a cyclic pursuit strategy. Automatica 43(8):1426–1431
11. Carli R, Chiuso A, Schenato L et al (2008) Distributed kalman filtering based on consensus strategies. IEEE J Sel Areas Commun 26(4):622–633

12. Wang JY, Liang HZ, Sun ZW et al (2012) Finite-time control for spacecraft formation with dual-number-based description. J Guid Control Dyn 35(3):950–962
13. Abdessameud A, Tayebi A (2011) Formation control of VTOL unmanned aerial vehicles with communication delays. Automatica 47(11):2383–2394
14. Wei CZ, Guo JF, Lu BG, et al (2013) Adaptive control for missile formation keeping under leader information unavailability. In: Proceedings of the 10th IEEE international conference on control and automation, pp 902–907
15. Cao YC, Stuart D, Ren W et al (2011) Distributed containment control for multiple autonomous vehicles with double-integrator dynamics: Algorithms and experiments. IEEE Trans Control Syst Technol 19(4):929–938
16. Kar S, Moura JMF (2010) Distributed consensus algorithms in sensor networks: Quantized data and random link failures. IEEE Trans Signal Process 58(3):1383–1400
17. Ray RJ, Cobleigh BR, Vachon MJ, et al (2002) Flight test techniques used to evaluate performance benefits during formation flight (Technical Report). NASA Dryden Research Center
18. Vitaliev V, Hayes J, Sharpe L (2012) Smarter skies the route to sustainability (News briefing). Eng Technol 7(9):20–21
19. Maithripala DHA, Jayasuriya S (2005) Radar deception through phantom track generation. In: Proceedings of the 2005 American control conference, pp 4102–4106
20. Kim J, Hespanha JP (2004) Cooperative radar jamming for groups of unmanned air vehicles. In: Proceedings of the 43rd IEEE conference on decision and control, pp 632–637
21. Sivakumar A, Tan CKY (2010) UAV swarm coordination using cooperative control for establishing a wireless communications backbone. In: Proceedings of the 9th international conference on autonomous agents and multiagent systems, pp 1157–1164
22. Burns R, McLaughlin A, Leitner J, et al (2000) TechSat 21: Formation design, control, and simulation. In: Proceedings of the 2000 IEEE aerospace conference, pp 19–25
23. Formation-flying satellite. http://www.wired.com/dangerroom/2013/05/formation-flying-satellites/
24. Alami R, Fleury S, Herrb M et al (1998) Multi-robot cooperation in the MARTHA project. IEEE Robot Autom Mag 5(1):36–47
25. Borkar V, Varaiya PP (1982) Asymptotic agreement in distributed estimation. IEEE Trans Autom Control 27(3):650–655
26. Tsitsiklis JN, Athans M (1984) Convergence and asymptotic agreement in distributed decision problems. IEEE Trans Autom Control 29(1):42–50
27. Vicsek T, Czirok A, Jacob EB et al (1995) Novel type of phase transitions in a system of self-driven particles. Phys Rev Lett 75(6):1226–1229
28. Jadbabaie A, Lin J, Morse AS (2003) Coordination of groups of mobile autonomous agents using nearest neighbor rules. IEEE Trans Autom Control 48(6):988–1001
29. Ren W, Beard RW (2005) Consensus seeking in multiagent systems under dynamically changing interaction topologies. IEEE Trans Autom Control 50(5):655–661
30. Lin P, Jia YM, Li L (2008) Distributed robust H_∞ consensus control in directed networks of agents with time-delay. Syst Control Lett 57(8):643–653
31. Bliman PA, Ferrari-Trecate G (2008) Average consensus problems in networks of agents with delayed communications. Automatica 44(8):1985–1995
32. Tian YP, Liu CL (2008) Consensus of multi-agent systems with diverse input and communication delays. IEEE Trans Autom Control 53(9):2122–2128
33. Xiao F, Wang L (2006) State consensus for multi-agent systems with switching topologies and time-varying delays. Int J Control 79(10):1277–1284
34. Sun YG, Wang L, Xie GM (2008) Average consensus in networks of dynamic agents with switching topologies and multiple time-varying delays. Syst Control Lett 57(2):175–183
35. Sun YG, Wang L (2009) Consensus of multi-agent systems in directed networks with nonuniform time-varying delays. IEEE Trans Autom Control 54(7):1607–1613
36. Hatano Y, Mesbahi M (2005) Agreement over random networks. IEEE Trans Autom Control 50(11):1867–1872

37. Porfiri M, Stilwell DJ (2007) Consensus seeking over random weighted directed graphs. IEEE Trans Autom Control 52(9):1767–1773
38. Cortés J (2006) Finite-time convergent gradient flows with applications to network consensus. Automatica 42(11):1993–2000
39. Xiao F, Wang L, Chen J et al (2009) Finite-time formation control for multi-agent systems. Automatica 45(11):2605–2611
40. Fang L. Antsaklis PJ (2005) Information consensus of asynchronous discrete-time multi-agent systems. In: Proceedings of the 2005 American control conference, pp 1883–1888
41. Mehyar M, Spanos D, Pongsajapan J et al (2007) Asynchronous distributed averaging on communication networks. IEEE-ACM Trans Netw 15(3):512–520
42. Xiao F, Wang L (2008) Asynchronous consensus in continuous-time multi-agent systems with switching topology and time-varying delays. IEEE Trans Autom Control 53(8):1804–1816
43. Ren W, Atkins E (2007) Distributed muti-vehicle coordinated control via local information exchange. Int J Robust Nonlinear Control 17(17):1002–1033
44. Xie GM, Wang L (2007) Consensus control for a class of networks of dynamic agents. Int J Robust Nonlinear Control 17(11):941–959
45. Lin P, Jia YM (2009) Further results on decentralised coordination in networks of agents with second-order dynamics. IET Contr Theory Appl 3(71):957–970
46. Tanner HG, Jadbabaie A, Pappas GJ (2007) Flocking in fixed and switching networks. IEEE Trans Autom Control 52(5):863–868
47. Lin P, Jia YM (2009) Consensus of second-order discrete-time multi-agent systems with nonuniform time-delays and dynamically changing topologies. Automatica 45(9):2154–2158
48. Lin P, Jia YM (2010) Consensus of a class of second-order multi-agent systems with time-delay and jointly-connected topologies. IEEE Trans Autom Control 55(3):778–784
49. Hu J, Lin YS (2010) Consensus control for multi-agent systems with double-integrator dynamics and time delays. IET Contr Theory Appl 4(1):109–118
50. Yu WW, Chen GR, Cao M (2010) Some necessary and sufficient conditions for second-order consensus in multi-agent dynamical systems. Automatica 46(6):1089–1095
51. Qin JH, Gao HJ, Zheng WX (2011) Second-order consensus for multi-agent systems with switching topology and communication delay. Syst Control Lett 60(6):390–397
52. Qin JH, Gao HJ, Zheng WX (2011) On average consensus in directed networks of agents with switching topology and time delay. Int J Syst Sci 42(12):1947–1956
53. Tian YP, Liu CL (2009) Robust consensus of multi-agent systems with diverse input delays and asymmetric interconnection perturbations. Automatica 45(5):1347–1353
54. Lin P, Jia YM (2010) Robust H_∞ consensus analysis of a class of second-order multi-agent systems with uncertainty. IET Contr Theory Appl 4(3):487–498
55. Zhang Y, Tian YP (2010) Consensus of data-sampled multi-agent systems with random communication delay and packet loss. IEEE Trans Autom Control 55(4):939–943
56. Yu WW, Zheng WX, Chen GR et al (2011) Second-order consensus in multi-agent dynamical systems with sampled position data. Automatica 47(7):1496–1503
57. Guan ZH, Meng C, Liao RQ et al (2012) Consensus of second-order multi-agent dynamic systems with quantized data. Phys Lett A 376(4):387–393
58. Yu WW, Chen GR, Cao M et al (2010) Second-order consensus for multiagent systems with directed topologies and nonlinear dynamics. IEEE Trans Syst Man Cybern Part B-Cybern 40(3):881–891
59. Munz U, Papachristodoulou A, Allgower F (2011) Robust consensus controller design for nonlinear relative degree two multi-agent systems with communication constraints. IEEE Trans Autom Control 56(1):145–151
60. Liu KE, Xie GM, Ren W et al (2013) Consensus for multi-agent systems with inherent nonlinear dynamics under directed topologies. Syst Control Lett 62(2):152–162
61. Xiao F, Wang L (2007) Consensus problems for high-dimensional multi-agent systems. IET Contr Theory Appl 1(3):830–837
62. Wang JH, Cheng DZ, Hu X (2008) Consensus of multi-agent linear dynamic systems. Asian J Control 10(1):144–155

63. Xi JX, Cai N, Zhong YS (2010) Consensus problems for high-order linear time-invariant swarm systems. Phys A 389(24):5619–5627
64. Ma CQ, Zhang JF (2010) Necessary and sufficient conditions for consensusability of linear multi-agent systems. IEEE Trans Autom Control 55(5):1263–1268
65. Li ZK, Duan ZS, Chen GR et al (2009) Consensus of multi-agent systems and synchronization of complex networks: a unified viewpoint. IEEE Trans Circuits Syst I-Regul Pap 57(1):213–224
66. Li ZK, Ren W, Liu XD et al (2013) Distributed consensus of linear multi-agent systems with adaptive dynamic protocols. Automatica 49(7):1986–1995
67. Seo JH, Shim H, Back J (2009) Consensus of high-order linear systems using dynamic output feedback compensator: Low gain approach. Automatica 45(11):2659–2664
68. Xi JX, Shi ZY, Zhong YS (2011) Consensus analysis and design for high-order linear swarm systems with time-varying delays. Physica A 390(23–24):4114–4123
69. Wang X, Saberi A, Stoorvogel AA et al (2013) Consensus in the network with uniform constant communication delay. Automatica 49(8):2461–2467
70. Scardovi L, Sepulchre R (2009) Synchronization in networks of identical linear systems. Automatica 45(11):2557–2562
71. Su YF, Huang J (2012) Stability of a class of linear switching systems with applications to two consensus problems. IEEE Trans Autom Control 57(6):1420–1430
72. Liu Y, Jia YM (2010) H_∞ consensus control of multi-agent systems with switching topology: a dynamic output feedback protocol. Int J Control 83(3):527–537
73. Zhao Y, Duan ZS, Wen GH et al (2012) Distributed H_∞ consensus of multi-agent systems: a performance region-based approach. Int J Control 85(3):332–341
74. Xi JX, Shi ZY, Zhong YS (2012) Output consensus analysis and design for high-order linear swarm systems: partial stability method. Automatica 48(9):2335–2343
75. Proskurnikov A (2013) Consensus in switching networks with sectorial nonlinear couplings: absolute stability approach. Automatica 49(2):488–495
76. Xi JX, Meng FL, Shi ZY et al (2012) Time-dependent admissible consensualization for singular time-delayed swarm systems. Syst Control Lett 61(11):1089–1096
77. Wang SM, Fu MY, Xu YJ, et al (2013) Consensus analysis of high-order singular multi-agent discrete-time systems. In: Proceedings of the 32nd Chinese control conference, pp 7136–7141
78. Kim H, Shim H, Seo JH (2011) Output consensus of heterogeneous uncertain linear multi-agent systems. IEEE Trans Autom Control 56(1):200–206
79. Tian YP, Zhang Y (2012) High-order consensus of heterogeneous multi-agent systems with unknown communication delays. Automatica 48(6):1205–1212
80. Chen YQ, Wang Z (2005) Formation control: A review and a new consideration. In: Proceedings of the 2005 IEEE/RSJ international conference on intelligent robots and systems, pp 3181–3186
81. Wang PKC (1991) Navigation strategies for multiple autonomous mobile robots moving in formation. J Robotic Syst 8(2):177–195
82. Balch T, Arkin R (1998) Behavior-based formation control for multirobot teams. IEEE Trans Robot Autom 14(6):926–939
83. Lewis M, Tan K (1997) High precision formation control of mobile robots using virtual structures. Auton Robot 4(4):387–403
84. Beard RW, Lawton J, Hadaegh FY (2001) A coordination architecture for spacecraft formation control. IEEE Trans Control Syst Technol 9(6):777–790
85. Ren W (2007) Consensus strategies for cooperative control of vehicle formations. IET Contr Theory Appl 1(2):505–512
86. Ren W, Sorensen N (2008) Distributed coordination architecture for multi-robot formation control. Robot Auton Syst 56(4):324–333
87. Xie GM, Wang L (2009) Moving formation convergence of a group of mobile robots via decentralised information feedback. Int J Syst Sci 40(10):1019–1027
88. Liu CL, Tian YP (2009) Formation control of multi-agent systems with heterogeneous communication delays. Int J Syst Sci 40(6):627–636

89. Lafferriere G, Williams A, Caughman J et al (2005) Decentralized control of vehicle formations. Syst Control Lett 54(9):899–910
90. Porfiri M, Roberson DG, Stilwell DJ (2007) Tracking and formation control of multiple autonomous agents: a two-level consensus approach. Automatica 43(8):1318–1328
91. Lin ZY, Francis B, Maggiore M (2005) Necessary and sufficient graphical conditions for formation control of unicycles. IEEE Trans Autom Control 50(1):121–127
92. Ma CQ, Zhang JF (2012) On formability of linear continuous-time multi-agent systems. J Syst Sci Complex 25(1):13–29
93. Dong XW, Xi JX, Lu G, et al (2013) Formation analysis and feasibility for high-order linear time-invariant swarm systems with time delays. In: Proceedings of the 32nd Chinese control conference, pp 7023–7029
94. Chung H, Sastry SS (2006) Autonomous helicopter formation using model predictive control. In: Proceedings of the 2006 AIAA guidance, Navigation, and Control conference, pp 459–473
95. Sharma RK, Ghose D (2009) Collision avoidance between UAV clusters using swarm intelligence techniques. Int J Syst Sci 40(5):521–538
96. Kim S, Kim Y (2009) Optimum design of three-dimensional behavioural decentralized controller for UAV formation flight. Eng Optimiz 41(3):199–224
97. Linorman NHM, Liu HHT (2008) Formation UAV flight control using virtual structure and motion synchronization. In: Proceedings of the 2008 American control conference, pp 1782–1787
98. Min HB, Sun FC, Niu F (2009) Decentralized UAV formation tracking flight control using gyroscopic force. In: Proceedings of international conference on computational intelligence for measurement systems and applications, pp 91–96
99. Qu Y, Zhu X, Zhang Y (2012) Cooperative control for UAV formation flight based on decentralized consensus algorithm. In: Intelligent robotics and applications lecture notes in computer science, pp 357–366
100. Gu Y, Campa G, Seanor B, et al (2009) Autonomous Formation Flight-Design and Experiments. In: Aerial vehicles, pp 233–256
101. Yun B, Chen BM, Lum KY et al (2010) Design and implementation of a leader-follower cooperative control system for unmanned helicopters. J Control Theor Appl 8(1):61–68
102. You DI, Shim DH (2011) Autonomous formation flight test of multi-micro aerial vehicles. J Intell Robot Syst 61(1–4):321–337
103. Di L, Chao H, Han J, et al (2011) Cognitive multi-UAV formation flight: Principle, low-cost UAV testbed, controller tuning and experiments. In: Proceedings of ASME/IEEE international conference on mechatronic and embedded systems and applications, DETC2011-47848
104. Kushleyev A, Mellinger D, Kumar V (2012) Towards a swarm of agile micro quadrotors. Robotics: Science and Systems
105. Turpin M, Michael N, Kumar V (2012) Trajectory design and control for aggressive formation flight with quadrotors. Auton Robot 33(1–2):143–156
106. Turpin M, Michael N, Kumar V (2012) Decentralized formation control with variable shapes for aerial robots. In: Proceedings of IEEE international conference on robotics and automation, pp 23–30
107. Yu BC, Dong XW, Shi ZY, et al (2013) Formation control for quadrotor swarm systems: Algorithms and experiments. In: Proceedings of the 32nd Chinese control conference, pp 7099–7104
108. Thornhill R, Alcock J (1983) The evolution of insect mating systems. Harvard University Press, Cambridge
109. Hummel HE, Miller TA (1984) Techniques in pheromone research. Springer, New York
110. Ren W (2007) Multi-vehicle consensus with a time-varying reference state. Syst Control Lett 56(7–8):474–483
111. Hong YG, Chen GR, Bushnell L (2008) Distributed observers design for leader-following control of multi-agent networks. Automatica 44(3):864–850
112. Ni W, Cheng DZ (2010) Leader-following consensus of multi-agent systems under fixed and switching topologies. Syst Control Lett 59(3–4):209–217

113. Li ZK, Liu XD, Ren W et al (2013) Distributed tracking control for linear multi-agent systems with a leader of bounded unknown input. IEEE Trans Autom Control 58(2):518–523
114. Ji M, Ferrari-Trecate G, Egerstedt M et al (2008) Containment control in mobile networks. IEEE Trans Autom Control 53(8):1972–1975
115. Meng ZY, Ren W, You Z (2010) Distributed finite-time attitude containment control for multiple rigid bodies. Automatica 46(12):2092–2099
116. Notarstefano G, Egerstedt M, Haque M (2011) Containment in leader-follower networks with switching communication topologies. Automatica 47(5):1035–1040
117. Cao YC, Ren W, Egerstedt M (2012) Distributed containment control with multiple stationary or dynamic leaders in fixed and switching directed networks. Automatica 48(8):1586–1597
118. Liu HY, Xie GM, Wang L (2012) Necessary and sufficient conditions for containment control of networked multi-agent systems. Automatica 48(7):1415–1422
119. Lou YC, Hong YG (2012) Target containment control of multi-agent systems with random switching interconnection topologies. Automatica 48(5):879–885
120. Liu HY, Xie GM, Wang L (2012) Containment of linear multi-agent systems under general interaction topologies. Syst Control Lett 61(4):528–534
121. Li ZK, Ren W, Liu XD et al (2013) Distributed containment control of multi-agent systems with general linear dynamics in the presence of multiple leaders. Int J Robust Nonlinear Control 23(5):534–547
122. Dong XW, Xi JX, Lu G et al (2014) Containment analysis and design for high-order linear time-invariant singular swarm systems with time delays. Int J Robust Nonlinear Control 24(7):1189–1204
123. Dong XW, Shi ZY, Lu G et al (2015) Output containment analysis and design for high-order linear time-invariant swarm systems. Int J Robust Nonlinear Control 25(6):900–913
124. Ferrari-Trecate G, Egerstedt M, Buffa A, et al (2006) Laplacian Sheep: A Hybrid, Stop-Go Policy for Leader-Based Containment Control. In: Proceedings of hybrid systems: computation and control, pp 212–226

Chapter 2
Preliminaries

Abstract This chapter introduces some basic definitions and results on graph theory, consensus decomposition of linear space theory, matrix theory, linear system theory, and singular system theory, which will be used in the following chapters. First, the definitions of directed graph, spanning tree, and Laplacian matrix, etc. are given, and properties of Laplacian matrix are addressed. Then the concepts of consensus subspace, complement consensus subspace, and state/output space decomposition are defined. Third, the properties of Kronecker product and Schur complement lemma are introduced. Moreover, the definitions and criteria for controllability, observability, and stability of linear time-invariant systems are summarized, and some results on partial stability of linear systems are also given. Finally, the definitions and results on the regularity, equivalent form, admissibility, and controllability of singular systems are introduced.

2.1 Graph Theory

A *graph* $G = (\mathscr{V}(G), \mathscr{E}(G))$ consists of a node set $\mathscr{V}(G) = \{v_1, v_2, \ldots, v_N\}$ and a edge set $\mathscr{E}(G) \subseteq \{(v_i, v_j), i \neq j;\ v_i, v_j \in \mathscr{V}(G)\}$. Denote by (v_i, v_j) and e_{ij} the edge from node v_i to node v_j, where v_i is called the *parent node* and v_j is called the *child node*. If a graph $G_0 = (\mathscr{V}_0(G_0), \mathscr{E}_0(G_0))$ satisfies $\mathscr{V}_0(G_0) \subseteq \mathscr{V}(G)$ and $\mathscr{E}_0(G_0) \subseteq \mathscr{E}(G)$, then G_0 is called a subgraph of G. If for any $e_{ij} \in \mathscr{E}(G)$, $e_{ji} \in \mathscr{E}(G)$, then G is an *undirected graph*, otherwise, G is a *directed graph*. A directed path from node v_{i_1} to v_{i_l} is a sequence of ordered edges with the form of $(v_{i_k}, v_{i_{k+1}})$, where $v_{i_k} \in \mathscr{V}(G)$ $(k = 1, 2, \ldots, l-1)$. For a directed graph G, if for any two different nodes v_i and v_j, there exists a directed path from node v_i to node v_j, then G is said to be strongly connected. If for any two different nodes v_i and v_j, there exists a node v_k that has directed paths to node v_i and v_j, then G is said to be weakly connected. For undirected graphs, weakly connected and strongly connected are equivalent, which can all be called as connected. A directed graph is said to have a *spanning tree* if there exists at least one node having a directed path to all the other nodes.

X. Dong, *Formation and Containment Control for High-order Linear Swarm Systems*, Springer Theses, DOI 10.1007/978-3-662-47836-3_2

Define the *adjacency matrix* of G as the nonnegative matrix $\mathscr{W} = [w_{ij}] \in \mathbb{R}^{N\times N}$ where w_{ij} represents the weight of edge e_{ji} with $w_{ij} > 0 \Leftrightarrow e_{ji} \in \mathscr{E}(G)$. Node v_i is called a *neighbor* of node v_j if there exists an edge e_{ij}. Denote by $N_i = \{v_j \in \mathscr{V}(G) : (v_j, v_i) \in \mathscr{E}(G)\}$ the neighbor set of node v_i. The in-degree and out-degree of node v_i are represented by $\deg_{in}(v_i) = \sum_{j=1}^{N} w_{ij}$ and $\deg_{out}(v_i) = \sum_{j=1}^{N} w_{ji}$. A graph G is balanced if for any node v_i, the in-degree is equal to the out-degree. Define the in-degree matrix of G by diagonal matrix $\mathscr{D}$ the elements of which are the in-degrees of nodes. The *Laplacian matrix* of G is defined as $L = \mathscr{D} - \mathscr{W}$.

The following lemmas show basic properties of the Laplacian matrix L.

Lemma 2.1 ([1, 2]) *For a directed graph G with N nodes, it holds that*

(i) L has at least one 0 eigenvalue, and $\mathbf{1}$ is the associated eigenvector; that is, $L\mathbf{1} = 0$;

(ii) If G has a spanning tree, then 0 is a simple eigenvalue of L, and all the other $N - 1$ eigenvalues have positive real parts;

(iii) If G does not have a spanning tree, then L has at least two 0 eigenvalues with the geometric multiplicity being not less than 2.

Lemma 2.2 ([3]) *For an undirected graph G with N nodes, it follows that*

(i) L has at least one 0 eigenvalue, and $\mathbf{1}$ is the associated eigenvector satisfying $L\mathbf{1} = 0$;

(ii) If G is connected, then 0 is a simple eigenvalue of L, and all the rest $N - 1$ eigenvalues are positive.

2.2 Consensus Decomposition of Linear Space

Define λ_i $(i = 1, 2, \ldots, N)$ as the eigenvalues of $L \in \mathbb{R}^N$, where the associated eigenvector of $\lambda_1 = 0$ is $\bar{u}_1 = \mathbf{1}$. Define a nonsingular matrix $U = [\bar{u}_1, \bar{u}_2, \ldots, \bar{u}_N] \in \mathbb{C}^{N\times N}$. Let $c_k \in \mathbb{R}^{\nu}$ $(k = 1, 2, \ldots, \nu)$ be linearly independent vectors and $p_j = \bar{u}_i \otimes c_k$ $(j = (i-1)\nu + k; i = 1, 2, \ldots, N; k = 1, 2, \ldots, \nu)$. A *consensus subspace* is defined as the subspace $\mathbb{C}(U)$ spanned by $p_k = \bar{u}_1 \otimes c_k = \mathbf{1} \otimes c_k$ $(k = 1, 2, \ldots, \nu)$, and a *complement consensus subspace* is defined as the subspace $\overline{\mathbb{C}}(U)$ spanned by $p_{\nu+1}, p_{\nu+2}, \ldots, p_{N\nu}$. Note that p_j $(j = 1, 2, \ldots, N\nu)$ are linearly independent. The following conclusion can be obtained.

Lemma 2.3 $\mathbb{C}(U) \oplus \overline{\mathbb{C}}(U) = \mathbb{C}^{N\nu}$.

Remark 2.1 From Lemma 2.3, one sees that any $\mathbb{C}^{N\nu}$ can be uniquely projected onto $\mathbb{C}(U)$ and $\overline{\mathbb{C}}(U)$. In the following chapters, the value of ν will be determined by the dimension of the state or output of each agent in the swarm systems. The decomposition of the state space or output space of the swarm system is called the *state space decomposition* or *output space decomposition*.

2.3 Matrix Theory

For matrices $A = [a_{ij}] \in \mathbb{R}^{m\times n}$ and $B \in \mathbb{R}^{p\times q}$, the *Kronecker product* can be defined as

$$A \otimes B = \begin{bmatrix} a_{11}B & a_{12}B & \cdots & a_{1n}B \\ a_{21}B & a_{22}B & \cdots & a_{2n}B \\ \vdots & \vdots & \ddots & \vdots \\ a_{m1}B & a_{m2}B & \cdots & a_{mn}B \end{bmatrix} \in \mathbb{R}^{(mp)\times(nq)};$$

and the direct sum is defined as

$$A \oplus B = \begin{bmatrix} A & 0 \\ 0 & B \end{bmatrix} \in \mathbb{R}^{(m+p)\times(n+q)}.$$

For matrices A, B, C, and D with appropriate dimensions, the Kronecker product has the following properties [4]:
(i) $A \otimes (B + C) = A \otimes B + A \otimes C$;
(ii) $(A \otimes B)(C \otimes D) = (AC) \otimes (BD)$;
(iii) $(A \otimes B)^T = A^T \otimes B^T$; and
(iv) $(A \otimes B)^{-1} = A^{-1} \otimes B^{-1}$.

Definition 2.1 For matrix $A \in \mathbb{C}^{n\times n}$, if all the eigenvalues of A have negative real parts, then A is called a *Hurwitz matrix* or stable matrix.

Lemma 2.4 (Schur complement [5]) *For a given matrix* $S = \begin{bmatrix} S_{11} & S_{12} \\ * & S_{22} \end{bmatrix}$, *where* $S_{11} \in \mathbb{R}^{r\times r}$, *the following statements are equivalent:*
(i) $S < 0$;
(ii) $S_{11} < 0$, $S_{22} - S_{12}^T S_{11}^{-1} S_{12} < 0$; *and*
(iii) $S_{22} < 0$, $S_{11} - S_{12} S_{22}^{-1} S_{12}^T < 0$.

2.4 Linear System Theory

Consider the following linear time-invariant system

$$\begin{cases} \dot{x}(t) = Ax(t) + Bu(t), \\ y(t) = Cx(t), \end{cases} \tag{2.1}$$

where $A \in \mathbb{R}^{n\times n}$, $B \in \mathbb{R}^{n\times m}$, $C \in \mathbb{R}^{q\times n}$, and $x(t) \in \mathbb{R}^n$, $u(t) \in \mathbb{R}^m$ and $y(t) \in \mathbb{R}^q$ is the state, control input and output, respectively.

Definition 2.2 For any initial state $x(0)$, if there exists control input $u(t)$ such that the state $x(t)$ of system (2.1) can converge to the origin in a finite time, then system (2.1) is called *controllable* or (A, B) is controllable.

Lemma 2.5 ([6]) *If* rank $\left[B, AB, \ldots, A^{n-1}B\right] = n$*, then* (A, B) *is controllable.*

Lemma 2.6 (Popov-Belevitch-Hautus (PBH) test for controllability [6]) *If* rank $[sI - A, B] = n$ $(\forall s \in \mathbb{C})$*, then* (A, B) *is controllable.*

Definition 2.3 If any initial state $x(0)$ of system (2.1) can be uniquely determined by the control input $u(t)$ and output $y(t)$ in a finite time, then system (2.1) is called *observable* or (C, A) is observable.

Lemma 2.7 ([6]) *If* rank$\left[C^T, A^T C^T, \ldots, (A^{n-1})^T C^T\right]^T = n$*, then* (C, A) *is observable.*

Lemma 2.8 (PBH test for observability [6]) *If* rank$\left[C^T, sI - A^T\right]^T = n$ $(\forall s \in \mathbb{C})$*, then* (C, A) *is observable.*

Definition 2.4 If matrix A is Hurwitz, then system (2.1) is *asymptotically stable.*

Lemma 2.9 *For system (2.1), the following statements are equivalent:*

(i) System (2.1) is asymptotically stable;

(ii) For any given positive matrix R*, the Lyapunov function* $A^T P + PA + R = 0$ *has positive definite solution* P*;*

(iii) There exists a positive definite matrix R *such that the Lyapunov function* $A^T P + PA + R = 0$ *has unique positive definite solution* P*; and*

(iv) There exists a positive definite matrix P *such that* $A^T P + PA < 0$.

Definition 2.5 If there exists a matrix $K \in \mathbb{R}^{m \times n}$ such that $A + BK$ is Hurwitz, then system (2.1) is *stabilizable* or (A, B) is stabilizable.

Lemma 2.10 ([7]) *System (2.1) is stabilizable if and only if* rank $[sI - A, B] = n$ $(\forall s \in \bar{\mathbb{C}}^+)$*, where* $\bar{\mathbb{C}}^+ = \{s | s \in \mathbb{C}, \mathrm{Re}(s) \geq 0\}$ *represents the closed right complex space.*

Definition 2.6 If there exists gain matrix $K \in \mathbb{R}^{n \times q}$ such that $A + KC$ is Hurwitz, then system (2.1) is *detectable* or (C, A) is detectable.

Lemma 2.11 ([7]) *System (2.1) is detectable if and only if* rank$\left[sI - A^T, C^T\right]^T = n$ $(\forall s \in \bar{\mathbb{C}}^+)$.

Consider the following linear time-invariant system

$$\begin{cases} \dot{x}(t) = Ax(t), \\ y(t) = Cx(t), \end{cases} \tag{2.2}$$

where $y(t) = \begin{bmatrix} y_o(t) \\ y_{\bar{o}}(t) \end{bmatrix}$, $A = \begin{bmatrix} A_{11} & A_{12} \\ A_{21} & A_{22} \end{bmatrix}$ and $C = [I, 0]$.

Definition 2.7 If for any given $\varepsilon > 0$, there exists $\delta = \delta(\varepsilon) > 0$ such that $\|y(0)\| < \delta \Rightarrow \|y_o(t)\| < \varepsilon$ ($\forall t \geq 0$), then system (2.2) is said to be stable with respect to $y_o(t)$.

Definition 2.8 If system (2.2) is stable with respect to $y_o(t)$ and $\lim_{t\to\infty} y_o(t) = 0$, then system (2.2) is said to be asymptotically stable with respect to $y_o(t)$.

Lemma 2.12 ([8]) *If (A_{22}, A_{12}) is completely observable, then system (2.2) is asymptotically stable with respect to $y_o(t)$ if and only if A is Hurwitz.*

If (A_{22}, A_{12}) is not completely observable, then there always exists a nonsingular matrix T such that

$$\left(T^{-1}A_{22}T, A_{12}T\right) = \left(\begin{bmatrix} D_1 & 0 \\ D_2 & D_3 \end{bmatrix}, [E_1, 0]\right), \quad T^{-1}A_{21} = \begin{bmatrix} F_1 \\ F_2 \end{bmatrix},$$

where (D_1, E_1) is completely observable. The following results can be obtained.

Lemma 2.13 ([8]) *If (A_{22}, A_{12}) is not completely observable, then system (2.2) is asymptotically stable with respect to $y_o(t)$ if and only if*

$$\begin{bmatrix} A_{11} & E_1 \\ F_1 & D_1 \end{bmatrix}$$

is Hurwitz.

2.5 Singular System Theory

Consider the following high-order LTI singular system

$$E\dot{x}(t) = Ax(t) + Bu(t), \tag{2.3}$$

where $A \in \mathbb{R}^{n\times n}$, $B \in \mathbb{R}^{n\times m}$, $E \in \mathbb{R}^{n\times n}$ satisfying $\mathrm{rank}(E) = r \leq n$, $x(t) \in \mathbb{R}^n$ is the state and $u_i(t) \in \mathbb{R}^m$ is the control input. In the following, the definitions and criteria for regularity, equivalent form, admissibility, and controllability of singular system (2.3) are summarized.

Definition 2.9 If there exists constant s_0 such that $\det(s_0E - A) \neq 0$, then system (2.3) is said to be *regular* or (E, A) is regular.

If (E, A) is regular, then there always exist nonsingular matrices P and Q such that

$$PEQ = \begin{bmatrix} I_r & 0 \\ 0 & N \end{bmatrix}, \quad PAQ = \begin{bmatrix} A_1 & 0 \\ 0 & I_{n-r} \end{bmatrix}, \quad PB = \begin{bmatrix} B_1 \\ B_2 \end{bmatrix},$$

where $N \in \mathbb{R}^{(n-r)\times(n-r)}$ represents the nilpotent matrix with nilpotent index l. Let $Q^{-1}x(t) = \left[x_1^T(t), x_2^T(t)\right]^T$. Then singular system (2.3) can be decomposed into

$$\dot{x}_1(t) = A_1 x_1(t) + B_1 u(t), \tag{2.4}$$

$$N\dot{x}_2(t) = x_2(t) + B_2 u(t). \tag{2.5}$$

This decomposition is often called as the *first equivalent form* of singular system (2.3) [9]. Subsystems (2.4) and (2.5) are said to be the slow subsystem and fast subsystem of singular system (2.3). Denote by $u^{(i)}(t)$ the ith derivative of $u(t)$. If the initial state $x(0)$ satisfies

$$x(0) = Q \begin{bmatrix} x_1(0) \\ -\sum_{i=0}^{l-1} N^i B_2 u^{(i)}(0) \end{bmatrix},$$

then $x(0)$ is said to be *admissible*.

Lemma 2.14 ([10]) *For given admissible initial state $x(0)$, singular system (2.3) has unique solution if and only if it is regular.*

Definition 2.10 If $\deg(\det(sE - A)) = \text{rank}(E)$ $(\forall s \in \mathbb{C})$, then singular system (2.3) is called *impulse-free* or (E, A) is impulse-free.

Lemma 2.15 ([10]) *Singular system (2.3) is impulse-free if and only if*

$$\text{rank} \begin{bmatrix} E & 0 \\ A & E \end{bmatrix} = n + \text{rank}(E).$$

Definition 2.11 If (E, A) is regular, impulse-free, and asymptotically stable, then singular system (2.3) is said to be admissible.

Definition 2.12 If for any $x_T \in \mathbb{R}^n$, $x(0) \in \mathbb{R}^n$ and $T > 0$, there exists admissible control input $u(t)$ such that $x(T) = x_T$, then singular system (2.3) is controllable or (E, A, B) is controllable.

Definition 2.13 If singular system (2.3) is controllable in $\mathbb{R}$, then it is called $\mathbb{R}$-*controllable*.

Lemma 2.16 ([11]) *Singular system (2.3) is $\mathbb{R}$-controllable if and only if the slow subsystem (2.4) is controllable or stabilizable.*

Lemma 2.17 ([11]) *Singular system (2.3) is controllable if and only if the slow subsystem (2.4) is controllable and*

$$\text{rank} \begin{bmatrix} E & 0 & 0 \\ A & E & B \end{bmatrix} = n + \text{rank}(E).$$

2.6 Conclusions

In this chapter, the basic definitions and results on graph theory, consensus decomposition of linear space theory, matrix theory, linear system theory, and singular system theory were introduced. These definitions and results are the research foundation of the following chapters.

References

1. Ren W, Beard RW (2005) Consensus seeking in multiagent systems under dynamically changing interaction topologies. IEEE Trans Autom Control 50(5):655–661
2. Xi JX, Cai N, Zhong YS (2010) Consensus problems for high-order linear time-invariant swarm systems. Phys A 389(24):5619–5627
3. Godsil C, Royle G (2001) Algebraic graph theory. Springer, New York
4. Horn RA, Johnson CR (1989) Topics in matrix analysis. Cambridge University Press, Cambridge
5. Boyd S, Ghaoui LE, Feron E et al (1994) Linear matrix inequalities in system and control theory. SIAM, Philadelphia
6. Williams RL, Lawrence DA (2007) Linear state-space control systems. Wiley, Hoboken
7. Chen CT (1999) Linear system theory and design. Oxford University Press, New York
8. Xi JX, Shi ZY, Zhong YS (2012) Output consensus analysis and design for high-order linear swarm systems: partial stability method. Automatica 48(9):2335–2343
9. Zhang QL, Liu C, Zhang X (2012) Complexity, analysis and control of singular biological systems. Springer, New York
10. Dai L (1989) Singular control systems. Springer, Berlin
11. Duan GR (2010) Analysis and design of descriptor linear systems. Springer, New York

Chapter 3
Consensus Control of Swarm Systems

Abstract This chapter studies practical consensus problems for general high-order linear time-invariant swarm systems with interaction uncertainties, nonuniform time-varying delays, and time-varying external disturbances on directed interaction topologies. A dynamic consensus protocol constructed by neighboring output information is adopted to deal with the practical consensus problem. Using the state space decomposition technique, practical consensus problems of swarm systems are converted into stability problems of disagreement subsystems. Based on the Lyapunov-Krasovskii functional approach and the linear matrix inequality technique, sufficient conditions for swarm systems to achieve practical consensus are proposed where the time-varying external disturbance can be in L_2 or L_∞. Explicit expressions of the practical consensus function and consensus error bounds are derived. Numerical simulations are presented to demonstrate the effectiveness of the obtained theoretical results.

3.1 Introduction

Consensus control is one of the most fundamental and important problems in cooperative control of swarm systems, which requires the states or outputs of all agents that reach an agreement. In [1–4], a special high-order LTI consensus model, which can be regarded as a controllability canonical form, was discussed. It should be mentioned that this special model can simplify the consensus problems significantly. In the general high-order LTI swarm systems, the dynamics of each agent has no specific structural characteristics compared with the dynamics in first-order, second-order, and special high-order LTI swarm systems, so the analysis for general high-order LTI swarm systems is more complicated and challenging. General high-order LTI models with fewer structural limitations were also considered by several researchers. In [5], the assumption that the consensus function is time-invariant was required. Swarm systems with time-varying consensus function were considered in [6, 7]. However, in [6–8], it was supposed that the interaction topology is fixed.

As is well known, in swarm systems, the interaction among agents may be time-varying and the time-varying interaction can be described by interaction uncertainties

X. Dong, *Formation and Containment Control for High-order Linear Swarm Systems*, Springer Theses, DOI 10.1007/978-3-662-47836-3_3

or switching interaction topologies. Interaction uncertainties were discussed in Lin and Jia [9], but the dynamics of each agent was of second-order and time-delays were not considered. The switching interaction topology was considered in [5, 10], where time-varying delays were not dealt with. Because of the movement of agents, the congestion of interaction channels and the asymmetry of interactions, etc., nonuniform time-varying delays may appear in swarm systems. Thus, considering the interaction uncertainties and nonuniform time-varying delays in general high-order LTI swarm systems is of practical importance. Moreover, time-varying external disturbances can arise in swarm systems due to actuator bias, calculation errors, severe environment, etc. In this case, it may be hard to achieve accurate consensus and previous results on accurate consensus may not work. To the best of our knowledge, practical consensus problems for general high-order LTI swarm systems with interaction uncertainties, nonuniform time-varying delays, and external disturbances are still open and challenging; here practical consensus means that all agents reach an agreement on certain variables of interest with certain error.

This chapter will mainly focus on discussing the above-mentioned practical consensus problem. First, a protocol based on the dynamic output feedback control is adopted. Then by model transformations, the original system is decomposed into a consensus subsystem and a disagreement subsystem. It is shown that the practical consensus problem of the original system is equivalent to the stability problem of the disagreement subsystem. Finally, by applying the Lyapunov-Krasovskii functional approach and the linear matrix inequality technique, sufficient conditions for swarm systems to achieve practical consensus are presented. Explicit expressions of the practical consensus function and the consensus error bound are also proposed.

The rest of this chapter is organized as follows. In Sect. 3.2, the problem description is presented. In Sect. 3.3, practical consensus problems of swarm systems are transformed into the stability problems of the disagreement subsystem. In Sect. 3.4, sufficient conditions for practical consensus problems are presented. Numerical simulations are given in Sect. 3.5. Finally, Sect. 3.6 concludes the whole work of this chapter.

3.2 Problem Description

Consider a high-order LTI swarm system with N agents. The interaction topology of the swarm system can be described by the directed graph G. Each agent can be regarded as a node in G. For $i, j \in \{1, 2, \ldots, N\}$, the available interaction from agent i to agent j can be denoted by edge e_{ij} and the interaction strength of e_{ij} can be denoted by the weight w_{ji}. Suppose that each agent has the general high-order LTI dynamics described by

$$\begin{cases} \dot{x}_i(t) = Ax_i(t) + Bu_i(t) + B_\omega \omega_i(t), \\ y_i(t) = Cx_i(t), \end{cases} \tag{3.1}$$

where $i \in \{1, 2, \ldots, N\}$, $x_i(t) \in \mathbb{R}^n$ is the state, $u_i(t) \in \mathbb{R}^m$ is the control input, $y_i(t) \in \mathbb{R}^q$ is the output, and $\omega_i(t) \in \mathbb{R}^{\bar{m}}$ is the external disturbance.

Consider the following dynamic output feedback protocol

$$\begin{cases} \dot{z}_i(t) = K_1 z_i(t) + K_2 \sum\limits_{j=1}^{N} (w_{ij} + \Delta w_{ij}(t))(y_j(t - \tau_{ij}(t)) - y_i(t - \tau_{ij}(t))), \\ u_i(t) = K_3 z_i(t) + K_4 \sum\limits_{j=1}^{N} (w_{ij} + \Delta w_{ij}(t))(y_j(t - \tau_{ij}(t)) - y_i(t - \tau_{ij}(t))), \end{cases} \tag{3.2}$$

where $i \in \{1, 2, \ldots, N\}$, $z_i(t) \in \mathbb{R}^{\bar{n}}$ is the state of the protocol, $K_1 \in \mathbb{R}^{\bar{n} \times \bar{n}}$, $K_2 \in \mathbb{R}^{\bar{n} \times q}$, $K_3 \in \mathbb{R}^{m \times \bar{n}}$ and $K_4 \in \mathbb{R}^{m \times q}$ are constant gain matrices, $\Delta w_{ij}(t)$ represents the uncertainty of the interaction strength, and $\tau_{ij}(t)$ is the time-varying delay from agent j to agent i. Let $\tau_r(t) \in \{\tau_{ij}(t) : i, j \in \{1, 2, \ldots, N\}\}$ $(r = 1, 2, \ldots, k)$, where k is the total number of time-delays. Assume $\tau_r(t) \in [0, \bar{\tau}_r]$ $(r = 1, 2, \ldots, k)$, where $\bar{\tau}_r$ stands for the upper bound of $\tau_r(t)$. Denote by d_r $(d_r < 1; r = 1, 2, \ldots, k)$ the upper bound of $|\dot{\tau}_r(t)|$. In practice, $\tau_r(t)$, $\bar{\tau}_r$ and d_r are mainly determined by the communication equipments and their values can be estimated according to the bandwidth of the communication channel, account of the communication data, transmit/receive time, and propagation delay [11]. Define $L_r = \left[l_{ij}^r\right] \in \mathbb{R}^{N \times N}$ and $\Delta L_r = \left[\Delta l_{ij}^r\right] \in \mathbb{R}^{N \times N}$, where

$$l_{ij}^r = \begin{cases} -w_{ij}, & i \neq j, \tau_{ij}(t) = \tau_r(t), \\ 0, & i \neq j, \tau_{ij}(t) \neq \tau_r(t), \\ -\sum_{s=1, s \neq i}^{N} l_{is}^r, & i = j, \end{cases}$$

$$\Delta l_{ij}^r = \begin{cases} -\Delta w_{ij}(t), & i \neq j, \tau_{ij}(t) = \tau_r(t), \\ 0, & i \neq j, \tau_{ij}(t) \neq \tau_r(t), \\ -\sum_{s=1, s \neq i}^{N} \Delta l_{is}^r(t), & i = j, \end{cases}$$

and ΔL_r describes the uncertainty of the interaction topology. Note that the interaction topology with uncertainty implies that the topology is time-varying. From the definition of L_r, it is easy to obtain that $L_r \mathbf{1} = 0$, $\Delta L_r \mathbf{1} = 0$, $\sum_{r=1}^{k} L_r = L$, and $\sum_{r=1}^{k} \Delta L_r = \Delta L$.

Let $\theta_i(t) = [x_i^T(t), z_i^T(t)]^T$ $(i = 1, 2, \ldots, N)$, $\theta(t) = [\theta_1^T(t), \theta_2^T(t), \ldots, \theta_N^T(t)]^T$, $\omega(t) = [\omega_1^T(t), \omega_2^T(t), \ldots, \omega_N^T(t)]^T$. Under protocol (3.2), the swarm system with the dynamics of each agent described by (3.1) can be written in a compact form as follows

$$\begin{cases} \dot{\theta}(t) = (I_N \otimes \bar{A})\theta(t) - \sum\limits_{r=1}^{k} \left((L_r + \Delta L_r) \otimes \bar{B}\right) \theta(t - \tau_r(t)) + (I_N \otimes \bar{B}_\omega)\omega(t), \quad t > 0, \\ \theta(t) = \varphi(t), \quad t \in \left[-\tilde{\tau}, 0\right], \end{cases} \tag{3.3}$$

where $\varphi(t)$ is a continuous vector-valued function on $t \in [-\tilde{\tau}, 0]$, $\tilde{\tau} = \max\{\bar{\tau}_1, \bar{\tau}_2, \ldots, \bar{\tau}_k\}$ and

$$\bar{A} = \begin{bmatrix} A & BK_3 \\ 0 & K_1 \end{bmatrix}, \ \bar{B} = \begin{bmatrix} BK_4C & 0 \\ K_2C & 0 \end{bmatrix}, \ \bar{B}_\omega = \begin{bmatrix} B_\omega \\ 0 \end{bmatrix}.$$

Definition 3.1 Swarm system (3.3) is said to *achieve state consensus* if for any given bounded initial states, there exists a vector-valued function $c(t) \in \mathbb{R}^{n+\bar{n}}$ dependent on the initial states such that $\lim_{t\to\infty}(\theta_i(t) - c(t)) = 0$ $(i = 1, 2, \ldots, N)$, where $c(t)$ is called a *state consensus function*.

If $\omega(t)$ does not vanish, under the influence of $\omega(t)$, it is difficult to achieve the accurate consensus. In view of this, we give a definition for state practical consensus.

Definition 3.2 Swarm system (3.3) is said to *achieve state practical consensus* if for any given bounded initial states, there exist a nonnegative constant δ and a vector-valued function $c(t) \in \mathbb{R}^{n+\bar{n}}$ dependent on the initial states such that $\lim_{t\to\infty} \|\theta_i(t) - c(t)\|_2 \leq \delta$ $(i = 1, 2, \ldots, N)$, where $c(t)$ and δ are called the *state practical consensus function* and *consensus error bound* respectively.

In the following, the problems that under what conditions swarm system (3.3) can achieve practical consensus are studied.

3.3 Problem Transformation and Preliminary Results

In this section, using state space decomposition, practical consensus problems of swarm systems are transformed into stability problems of the disagreement subsystem and preliminary results including the explicit expression of the state practical consensus function are presented.

Define nonsingular matrices $\bar{U} = [\bar{u}_2, \ldots, \bar{u}_N]$ and $U^{-1} = [\tilde{u}_1, \tilde{U}^H]^H$. Since $(L_r + \Delta L_r)\mathbf{1} = 0$ $(r = 1, 2, \ldots, k)$, one can obtain

$$U^{-1}(L_r+\Delta L_r)U = \begin{bmatrix} \tilde{u}_1^H \\ \tilde{U} \end{bmatrix}(L_r+\Delta L_r)\left[\bar{u}_1, \bar{U}\right] = \begin{bmatrix} 0 & \tilde{u}_1^H(L_r + \Delta L_r)\bar{U} \\ 0 & \tilde{U}(L_r + \Delta L_r)\bar{U} \end{bmatrix}. \quad (3.4)$$

Let $\tilde{\theta}(t) = (U^{-1} \otimes I_{n+\bar{n}})\theta(t) = \left[\tilde{\theta}_1^H(t), \tilde{\theta}_2^H(t), \ldots, \tilde{\theta}_N^H(t)\right]^H$. By (3.4), swarm system (3.3) can be transformed into

$$\begin{aligned}\dot{\tilde{\theta}}(t) = (I_N \otimes \bar{A})\tilde{\theta}(t) &- \sum_{r=1}^{k}\left(\begin{bmatrix} 0 & \tilde{u}_1^H(L_r + \Delta L_r)\bar{U} \\ 0 & \tilde{U}(L_r + \Delta L_r)\bar{U} \end{bmatrix} \otimes \bar{B}\right)\tilde{\theta}(t - \tau_r(t)) \\ &+ \left(\begin{bmatrix} \tilde{u}_1^H \\ \tilde{U} \end{bmatrix} \otimes \bar{B}_\omega\right)\omega(t). \end{aligned} \quad (3.5)$$

Define $\varsigma_p(t) = \left[\tilde{\theta}_2^H(t), \ldots, \tilde{\theta}_N^H(t)\right]^H$, then system (3.5) can be rewritten as

$$\dot{\tilde{\theta}}_1(t) = \bar{A}\tilde{\theta}_1(t) - \sum_{r=1}^{k} \left(\tilde{u}_1^H(L_r + \Delta L_r)\bar{U} \otimes \bar{B}\right)\varsigma_p(t - \tau_r(t)) + (\tilde{u}_1^H \otimes \bar{B}_\omega)\omega(t), \tag{3.6}$$

$$\begin{aligned}\dot{\varsigma}_p(t) &= (I_{N-1} \otimes \bar{A})\varsigma_p(t) - \sum_{r=1}^{k} \left(\tilde{U}(L_r + \Delta L_r)\bar{U} \otimes \bar{B}\right)\varsigma_p(t - \tau_r(t)) \\ &\quad + (\tilde{U} \otimes \bar{B}_\omega)\omega(t).\end{aligned} \tag{3.7}$$

Choose the ν in Sect. 2.2 as $\nu = n + \bar{n}$. From Lemma 2.3, one has that the state of swarm system (3.3) can be uniquely projected onto $\mathbb{C}(U)$ and $\bar{\mathbb{C}}(U)$; that is, there exist $\alpha_j(t)$ $(j = 1, 2, \ldots, N(n + \bar{n}))$ such that $\theta(t) = \theta_C(t) + \theta_{\bar{C}}(t)$, where $\theta_C(t) = \sum_{j=1}^{n+\bar{n}} \alpha_j(t)p_j$ and $\theta_{\bar{C}}(t) = \sum_{j=n+\bar{n}+1}^{N(n+\bar{n})} \alpha_j(t)p_j$. Note that $\varsigma_p(t) = \left[0, I_{(N-1)(n+\bar{n})}\right](U^{-1} \otimes I_{n+\bar{n}})\theta(t), \tilde{\theta}_1(t) = \left[I_{n+\bar{n}}, 0\right](U^{-1} \otimes I_{n+\bar{n}})\theta(t), \bar{u}_i \otimes c_j = (U \otimes I_{n+\bar{n}})(e_i \otimes c_j)$, where $i \in \{1, 2, \ldots, N\}$, $j \in \{1, 2, \ldots, n + \bar{n}\}$ and $e_i \in \mathbb{R}^N$ has a 1 as its ith entry and 0 elsewhere. It follows that

$$\theta_{\bar{C}}(t) = (U \otimes I_{n+\bar{n}})\left[0, \varsigma_p^H(t)\right]^H, \tag{3.8}$$

$$\theta_C(t) = (U \otimes I_{n+\bar{n}})\left[\tilde{\theta}_1^H(t), 0\right]^H = (U \otimes I_{n+\bar{n}})\left(e_1 \otimes \tilde{\theta}_1(t)\right) = \mathbf{1} \otimes \tilde{\theta}_1(t). \tag{3.9}$$

Based on the structures of p_j $(j = 1, 2, \ldots, N(n + \bar{n}))$, the subsystems with states denoted by $\theta_C(t)$ and $\theta_{\bar{C}}(t)$ describe the consensus dynamics and disagreement dynamics of swarm system (3.3), respectively. By (3.5)–(3.9), the following lemmas can be obtained.

Lemma 3.1 *For the given protocol (3.2), swarm system (3.3) with $\omega(t) \equiv 0$ achieves state consensus if and only if subsystem (3.7) is asymptotically stable.*

Subsystem (3.7) represents the disagreement component of the original swarm system. If subsystem (3.7) is asymptotically stable, the disagreement component vanishes and only the consensus component remains, which means that $\lim_{t\to\infty}(\theta_i(t) - \tilde{\theta}_1(t)) = 0$ $(i = 1, 2, \ldots, N)$. This is the reason why subspaces $\mathbb{C}(U)$ and $\bar{\mathbb{C}}(U)$ in Sect. 2.2 are called consensus subspace and complement consensus subspace. On the other hand, if the disagreement component does not vanish ultimately but keeps a bounded value for bounded input, then there exist a consensus component and a bounded disagreement component simultaneously. In this case, the following lemma is present.

Lemma 3.2 *For the given protocol (3.2), swarm system (3.3) achieves state practical consensus if and only if subsystem (3.7) is bounded input and bounded state (BIBS) stable.*

Lemmas 3.1 and 3.2 provide criteria for checking whether a swarm system achieves state consensus or state practical consensus. It can be seen that Lemma 3.1 is a special case of Lemma 3.2. If swarm system (3.3) achieves practical consensus, one can obtain the expression of the state practical consensus function by solving the differential equation (3.6).

Lemma 3.3 *If swarm system (3.3) achieves state practical consensus, then its state practical consensus function $c(t)$ satisfies*

$$\lim_{t\to\infty}(c(t) - (c_0(t) - c_\tau(t) - c_\Delta(t)) - c_\omega(t)) = 0,$$

where

$$c_0(t) = e^{\bar{A}t}[I_{n+\bar{n}}, 0, \ldots, 0]P_{\mathbb{C}(U),\bar{\mathbb{C}}(U)}\theta(0) - \int_0^t \left(e^{\bar{A}(t-s)}\tilde{u}_1^H L\bar{U} \otimes \bar{B}\varsigma_p(s)\right)ds,$$

$$c_\tau(t) = \sum_{r=1}^{k}\int_0^t \left(e^{\bar{A}(t-s)}\tilde{u}_1^H L_r\bar{U} \otimes \bar{B}(\varsigma_p(t-\tau_r(s)) - \varsigma_p(s))\right)ds,$$

$$c_\Delta(t) = \sum_{r=1}^{k}\int_0^t \left(e^{\bar{A}(t-s)}\tilde{u}_1^H \Delta L_r\bar{U} \otimes \bar{B}\varsigma_p(t-\tau_r(s))\right)ds,$$

$$c_\omega(t) = \int_0^t \left(e^{\bar{A}(t-s)}\tilde{U} \otimes \bar{B}_\omega\omega(s)\right)ds,$$

with $P_{\mathbb{C}(U),\bar{\mathbb{C}}(U)} = [p_1, \ldots, p_{n+\bar{n}}, 0, \ldots, 0]P^{-1}$ being an oblique projector onto $\mathbb{C}(U)$ along $\bar{\mathbb{C}}(U)$ and $P = [p_1, p_2, \ldots, p_{N(n+\bar{n})}]$.

In Lemma 3.3, $c_0(t)$ is said to be the nominal function which describes the consensus function of the swarm system without time-varying delays, interaction uncertainties, and external disturbances. $c_\tau(t)$, $c_\Delta(t)$, and $c_\omega(t)$ describe the impacts of time-varying delays, interaction uncertainties, and external disturbances, respectively.

Remark 3.1 An explicit expression of the state practical consensus function for swarm system (3.3) is presented and the impacts of time-varying delays, interaction uncertainties, and external disturbances are described in Lemma 3.3. In Olfati-Saber and Murray [12], the χ-consensus problem, which determines the consensus function, was discussed. It was shown that if the interaction topology of a swarm system is balanced and strongly connected, then the consensus function is the average value of the initial states of all agents. But when interaction topologies are general directed, and time-varying delays, interaction uncertainties, and external disturbances are considered, to the best of our knowledge, there is not a general method shown in the literature to determine the consensus function.

If the Laplacian matrix is balanced, then $\mathbf{1}^T L = 0$. Since U consists of eigenvectors and generalized eigenvectors of L, it can be shown that $\mathbf{1}^T \bar{U} = 0$. In this case, one has $\tilde{u}_1^T = \mathbf{1}^T / N$, $\tilde{u}_1^T L = 0$, $\tilde{u}_1^T L_r = 0$ and $\tilde{u}_1^T \Delta L_r = 0$. Let $\tilde{C} = \left[c_1, c_2, \ldots, c_{n+\bar{n}}\right]$. It holds that

$$P = U \otimes \tilde{C},$$

$$[p_1, \ldots, p_{n+\bar{n}}, 0, \ldots, 0] = [\bar{u}_1, 0] \otimes \tilde{C}.$$

Therefore,

$$P_{\mathbb{C}(U),\bar{\mathbb{C}}(U)} = \left([\bar{u}_1, 0] \otimes \tilde{C}\right)\left(U^{-1} \otimes \tilde{C}^{-1}\right) = \left([\,\tfrac{1}{N} \;\cdots\; \tfrac{1}{N}\,]^T \otimes I_{n+\bar{n}}\right).$$

Then by Lemma 3.3, the following corollary can be obtained.

Corollary 3.1 *If swarm system (3.3) achieves state practical consensus and the topology is balanced, then its state practical consensus function $c(t)$ satisfies*

$$\lim_{t\to\infty} (c(t) - c_0(t) - c_\omega(t)) = 0,$$

where

$$c_0(t) = \frac{1}{N} e^{\bar{A}t} \sum_{i=1}^{N} \theta_i(0),$$

$$c_\omega(t) = \int_0^t \left(e^{\bar{A}(t-s)} \tilde{U} \otimes \bar{B}_\omega \omega(s)\right) ds.$$

Remark 3.2 For first-order swarm systems, Olfati-Saber and Murry [12] proposed the average consensus problem; that is, the consensus function is the average of initial states of all agents. In Corollary 3.1, if $\bar{A} = 0$ and $\omega \equiv 0$, then swarm system (3.3) becomes a typical first-order swarm system and $c(t) = \sum_{i=1}^{N} \theta_i(0) \big/ N$. In this case, $\delta = 0$; that is, accurate state consensus can be achieved. One can see that average consensus is a special case of ours. Moreover, for swarm system (3.3) with balanced topologies, Corollary 3.1 presents a general approach to determine the consensus function $c(t)$ which can be regarded as the practical average consensus function.

3.4 State Practical Consensus Analysis

In this section, first, a necessary and sufficient condition for swarm system (3.3) with $\omega(t) \in L_2$, $\Delta L_r \equiv 0$ and $\tau_r \equiv 0$ $(r = 1, 2, \ldots, k)$ to achieve state practical consensus is presented. Then by applying the Lyapunov-Krasovskii functional approach and the LMI technique, sufficient conditions for swarm system (3.3) to achieve state practical consensus are proposed. For simplicity of description, the results are given in two theorems. The first theorem gives the results for system (3.3) with $\Delta L_r \equiv 0$ $(r = 1, 2, \ldots, k)$ and $\omega(t) \equiv 0$ to achieve state consensus. Finally, sufficient conditions for swarm system (3.3) to achieve state practical consensus are proposed and the consensus error bound is derived.

By Lemma 4.1 in [13] and Lemma 3.2, the following lemma can be obtained directly.

Lemma 3.4 *For a given positive constant* μ_0*, swarm system (3.3) with* $\omega(t) \in L_2$*,* $\Delta L_r \equiv 0$ *and* $\tau_r(t) \equiv 0$ $(r = 1, 2, \ldots, k)$ *achieves state practical consensus with* $\int_0^T \varsigma_p^H(t)\varsigma_p(t)d(t) \leqslant \mu_0 \int_0^T \omega^T(t)\omega(t)d(t)$*, if and only if there exists a positive symmetric matrix* P*, such that the following LMI is feasible*

$$\begin{bmatrix} \left(I_{N-1} \otimes \bar{A} - \tilde{U}L\bar{U} \otimes \bar{B}\right)^T P + P\left(I_{N-1} \otimes \bar{A} - \tilde{U}L\bar{U} \otimes \bar{B}\right) & P(\tilde{U} \otimes \bar{B}_\omega) & I \\ * & -\mu_0 I & 0 \\ * & * & -\mu_0 I \end{bmatrix} < 0.$$

Theorem 3.1 *Swarm system (3.3) with* $\Delta L_r \equiv 0$ $(r = 1, 2, \ldots, k)$ *and* $\omega(t) \equiv 0$ *achieves state consensus if there exist real symmetric matrices* $R > 0$*,* $Q_r > 0$*,* $S_r > 0$*,* $M_r = \begin{bmatrix} M_{r,11} & M_{r,12} \\ * & M_{r,22} \end{bmatrix} \geq 0$*,* X_r*,* Y_r*, such that the following LMIs are feasible*

$$\Xi = \begin{bmatrix} \Xi_{11} & \Xi_{12} & \Xi_{13} \\ * & \Xi_{22} & \Xi_{23} \\ * & * & \Xi_{33} \end{bmatrix} < 0, \tag{3.10}$$

$$\Upsilon_r = \begin{bmatrix} M_{r,11} & M_{r,12} & X_r \\ * & M_{r,22} & Y_r \\ * & * & S_r \end{bmatrix} \geq 0, \tag{3.11}$$

where $r = 1, 2, \ldots, k$ *and*

$$\Xi_{11} = R(I_{N-1} \otimes \bar{A}) + (I_{N-1} \otimes \bar{A})^T R + \sum_{r=1}^{k} Q_r + \sum_{r=1}^{k} (X_r + X_r^T) + \sum_{r=1}^{k} \bar{\tau}_r M_{r,11},$$

$$\Xi_{12} = \left[-R\left(\tilde{U}L_1\bar{U} \otimes \bar{B}\right) - X_1 + Y_1^T + \bar{\tau}_1 M_{1,12}, \ldots, -R\left(\tilde{U}L_k\bar{U} \otimes \bar{B}\right) - X_k + Y_k^T + \bar{\tau}_k M_{k,12}\right],$$

$$\Xi_{13} = \left[\bar{\tau}_1(I_{N-1} \otimes \bar{A})^T S_1, \ldots, \bar{\tau}_k(I_{N-1} \otimes \bar{A})^T S_k\right],$$

$$\Xi_{22} = \text{blockdiag}\left\{(d_1 - 1)Q_1 - Y_1 - Y_1^T + \bar{\tau}_1 M_{1,22}, \ldots, (d_k - 1)Q_k - Y_k - Y_k^T + \bar{\tau}_k M_{k,22}\right\},$$

$$\Xi_{23} = \begin{bmatrix} -\bar{\tau}_1 \left(\tilde{U} L_1 \bar{U} \otimes \bar{B}\right)^T S_1 & \ldots & -\bar{\tau}_k \left(\tilde{U} L_1 \bar{U} \otimes \bar{B}\right)^T S_k \\ \vdots & \ddots & \vdots \\ -\bar{\tau}_1 \left(\tilde{U} L_k \bar{U} \otimes \bar{B}\right)^T S_1 & \ldots & -\bar{\tau}_k \left(\tilde{U} L_k \bar{U} \otimes \bar{B}\right)^T S_k \end{bmatrix},$$

$$\Xi_{33} = \text{blockdiag}\{-\bar{\tau}_1 S_1, \ldots, -\bar{\tau}_k S_k\}.$$

Proof Consider the following Lyapunov-Krasovskii functional candidate

$$V(t) = V_1(t) + V_2(t) + V_3(t), \tag{3.12}$$

where

$$V_1(t) = \varsigma_p^H(t) R \varsigma_p(t),$$

$$V_2(t) = \sum_{r=1}^{k} \int_{t-\tau_r(t)}^{t} \varsigma_p^H(\alpha) Q_r \varsigma_p(\alpha) d\alpha,$$

$$V_3(t) = \sum_{r=1}^{k} \int_{-\bar{\tau}_r}^{0} \int_{t+\alpha}^{t} \dot{\varsigma}_p^H(s) S_r \dot{\varsigma}_p(s) ds d\alpha.$$

Taking the time derivative of $V(t)$ along the trajectory of subsystem (3.7) yields

$$\begin{aligned} \dot{V}_1(t) = {} & \varsigma_p^H(t)(R(I_{N-1} \otimes A) + (I_{N-1} \otimes A)^T R)\varsigma_p(t) \\ & - \sum_{r=1}^{k} 2\varsigma_p^H(t) R \left(\tilde{U} L_r \bar{U} \otimes \bar{B}\right) \varsigma_p(t - \tau_r(t)), \end{aligned} \tag{3.13}$$

$$\dot{V}_2(t) \le \sum_{r=1}^{k} \varsigma_p^H(t) Q_r \varsigma_p(t) - \sum_{r=1}^{k} (1 - d_r) \varsigma_p^H(t - \tau_r(t)) Q_r \varsigma_p(t - \tau_r(t)), \tag{3.14}$$

$$\dot{V}_3(t) \le \sum_{r=1}^{k} \bar{\tau}_r \dot{\varsigma}_p^H(t) S_r \dot{\varsigma}_p(t) - \sum_{r=1}^{k} \int_{t-\tau_r(t)}^{t} \dot{\varsigma}_p^H(\alpha) S_r \dot{\varsigma}_p(\alpha) d\alpha, \tag{3.15}$$

where

$$\sum_{r=1}^{k} \bar{\tau}_r \dot{\varsigma}_p^H(t) S_r \dot{\varsigma}_p(t) = \sum_{r=1}^{k} \bar{\tau}_r \xi_0^H(t) \Big[I_{N-1} \otimes \bar{A}, -\tilde{U} L_1 \bar{U} \otimes \bar{B}, \ldots, -\tilde{U} L_k \bar{U} \otimes \bar{B} \Big]^T \\ \times S_r \Big[I_{N-1} \otimes \bar{A}, -\tilde{U} L_1 \bar{U} \otimes \bar{B}, \ldots, -\tilde{U} L_k \bar{U} \otimes \bar{B} \Big] \xi_0(t),$$

with $\xi_0(t) = \Big[\varsigma_p^H(t), \varsigma_p^H(t - \tau_1(t)), \ldots, \varsigma_p^H(t - \tau_k(t)) \Big]^H$. Since

$$\int_{t-\tau_r(t)}^{t} \dot{\varsigma}_p(\alpha) d\alpha = \varsigma_p(t) - \varsigma_p(t - \tau_r(t)),$$

for any appropriately dimensioned real matrices X_r and Y_r $(r = 1, 2, \ldots, k)$, one obtains

$$\Gamma_1 = \sum_{r=1}^{k} 2 \Big[\varsigma_p^H(t), \varsigma_p^H(t - \tau_r(t)) \Big] \begin{bmatrix} X_r \\ Y_r \end{bmatrix} (\varsigma_p(t) - \varsigma_p(t - \tau_r(t)) - \textstyle\int_{t-\tau_r(t)}^{t} \dot{\varsigma}_p(\alpha) d\alpha) = 0. \tag{3.16}$$

In addition, for any real symmetric matrices $M_r = \begin{bmatrix} M_{r,11} & M_{r,12} \\ * & M_{r,22} \end{bmatrix} \geq 0$ $(r = 1, 2, \ldots, k)$, the following holds

$$\Gamma_2 = \sum_{r=1}^{k} \bar{\tau}_r \Big[\varsigma_p^H(t), \varsigma_p^H(t - \tau_r(t)) \Big] M_r \begin{bmatrix} \varsigma_p(t) \\ \varsigma_p(t - \tau_r(t)) \end{bmatrix} \\ - \sum_{r=1}^{k} \textstyle\int_{t-\tau_r(t)}^{t} \Big[\varsigma_p^H(t), \varsigma_p^H(t - \tau_r(t)) \Big] M_r \begin{bmatrix} \varsigma_p(t) \\ \varsigma_p(t - \tau_r(t)) \end{bmatrix} d\alpha \geq 0. \tag{3.17}$$

By (3.13)–(3.17), it holds that

$$\dot{V}(t) \leq \dot{V}_1(t) + \dot{V}_2(t) + \dot{V}_3(t) + \Gamma_1 + \Gamma_2 \leq \xi_0^H(t) \Xi_0 \xi_0(t) \\ - \sum_{r=1}^{k} \int_{t-\tau_r(t)}^{t} \xi_r^H(t, \alpha) \Upsilon_r \xi_r(t, \alpha) d\alpha, \tag{3.18}$$

where $\xi_r(t, \alpha) = [\varsigma_p^H(t), \varsigma_p^H(t - \tau_r(t)), \dot{\varsigma}_p^H(\alpha)]^H$ $(r = 1, 2, \ldots, k)$,

$$\Xi_0 = \begin{bmatrix} \Xi_{11} & \Xi_{12} \\ * & \Xi_{22} \end{bmatrix} + \sum_{r=1}^{k} \bar{\tau}_r \Big[I_{N-1} \otimes \bar{A}, -\tilde{U} L_1 \bar{U} \otimes \bar{B}, \ldots, -\tilde{U} L_k \bar{U} \otimes \bar{B} \Big]^T \\ \times S_r \Big[I_{N-1} \otimes \bar{A}, -\tilde{U} L_1 \bar{U} \otimes \bar{B}, \ldots, -\tilde{U} L_k \bar{U} \otimes \bar{B} \Big], \tag{3.19}$$

with Υ_r $(r = 1, 2, \ldots, k)$ given in (3.11).

From (3.18), one knows that if $\Xi_0 < 0$ and $\Upsilon_r \geq 0$ $(r = 1, 2, \ldots, k)$, then $\dot{V}(t) < -\upsilon_0 \left\| \varsigma_p(t) \right\|_2^2$ for a positive constant υ, which means that subsystem (3.7) is asymptotically stable. By Lemma 3.1, swarm system (3.3) with $\omega(t) \equiv 0$ and $\Delta L_r \equiv 0$ achieves state consensus. Using Schur complement, one gets that $\Xi_0 < 0$ is equivalent to LMI (3.10). This completes the proof.

Before moving on, two lemmas need to be introduced for analyzing the conditions for swarm system (3.3) to achieve practical consensus.

Lemma 3.5 *Define G_r the subgraph of G corresponding to L_r. Let $\mathscr{D}_r$ be the* 0-1 *matrix with rows and columns indexed by the nodes and edges of G, and E_r be the* 0-1 *matrix with rows and columns indexed by the edges and nodes of G, such that*

$$\mathscr{D}_{ruf} = \begin{cases} 1, & if\ the\ node\ u\ is\ the\ child\ node\ of\ the\ edge\ f, \\ 0, & otherwise, \end{cases}$$

$$\mathscr{E}_{rfu} = \begin{cases} 1, & if\ the\ node\ u\ is\ the\ parent\ node\ of\ the\ edge\ f, \\ 0, & otherwise. \end{cases}$$

Let $\mathscr{D} = \sum_{r=1}^{k} \mathscr{D}_r$ and $\Lambda = \operatorname{diag}\{\mu_1, \mu_2, \ldots, \mu_\iota\}$, where μ_p $(p = 1, 2, \ldots, \iota)$ is the weight of the pth edge of G and ι is the number of the edges of G. Then L_r can be denoted by $L_r = \mathscr{D}\Lambda(\mathscr{D}_r^T - \mathscr{E}_r)$ $(r = 1, 2, \ldots, k)$.

Proof For any $r \in \{1, 2, \ldots, k\}$, the (i, j)th element of $\mathscr{D}\Lambda\mathscr{D}_r{}^T$ can be written as $\sum_{l=1}^{\iota} \mathscr{D}_{il}\mu_l\mathscr{D}_{rjl}$. Since $L = \sum_{r=1}^{k} L_r$, G has the same nodes as G_r $(r = 1, 2, \ldots, k)$, and the edges of G consist of the ones of G_r $(r = 1, 2, \ldots, k)$ without any superposition. Because any edge only has one parent node, it follows that $\mathscr{D}_{il}\mathscr{D}_{rjl} = 0$ for any $i \neq j$. Thus one has that $\mathscr{D}\Lambda\mathscr{D}_r{}^T$ is a diagonal matrix with the (i, i)th element equal to the in-degree of the node v_i of G_r. Hence the degree matrix of G_r satisfies $\tilde{\mathscr{D}}_r = \mathscr{D}\Lambda\mathscr{D}_r{}^T$. Similarly, the (i, j)th element of $\mathscr{D}\Lambda\mathscr{E}_r$ can be written as $\sum_{l=1}^{\iota} \mathscr{D}_{il}\mu_l\mathscr{E}_{rlj}$, which is equal to the weight of the edge (v_j, v_i) of G_r; therefore the adjacency matrix of G_r can be denoted by $\tilde{\mathscr{W}}_r = \mathscr{D}\Lambda\mathscr{E}_r$. Since $L_r = \tilde{\mathscr{D}}_r - \tilde{\mathscr{W}}_r$ by the definition of the Laplacian matrix in Sect. 2.1, one has $L_r = \mathscr{D}\Lambda(\mathscr{D}_r{}^T - \mathscr{E}_r)$. By considering all subgraphs G_r $(r=1, 2, \ldots, k)$, the conclusion of Lemma 3.5 is obtained.

By Lemma 3.5, the uncertainty matrix ΔL_r of L_r can be written as

$$\Delta L_r = \mathscr{D}\bar{\Lambda}(t)\bar{\mathscr{E}}_r,$$

where $\bar{\mathscr{E}}_r \in \mathbb{R}^{\iota \times N}$ and $\bar{\Lambda}(t)$ is a diagonal matrix whose diagonal elements are uncertainties of the interaction strength. Since $\Delta L = \sum_{r=1}^{k} \Delta L_r$, one has $\Delta L = \mathscr{D}\bar{\Lambda}(t)\sum_{r=1}^{k}\bar{\mathscr{E}}_r$. Assume $\left|\Delta w_{ij}(t)\right| / w_{ij} \leq 1$ $(i, j \in \{1, 2, \ldots, N\})$. Without loss of generality, it is assumed that $\bar{\Lambda}^T(t)\bar{\Lambda}(t) \leq I_\iota$ $(\forall t)$.

Lemma 3.6 ([14]) *Given matrices H, K and a positive scalar γ, for all $\bar{\Lambda}(t)$ satisfying $\bar{\Lambda}^T(t)\bar{\Lambda}(t) \leq I$, the following inequality holds*

$$H\bar{\Lambda}(t)K + K^T\bar{\Lambda}^T(t)H^T \leq \gamma^{-1}HH^T + \gamma K^T K.$$

Sufficient conditions for swarm system (3.3) to achieve state practical consensus are given as follows.

Theorem 3.2 *Swarm system (3.3) achieves state practical consensus for any $\omega(t) \in L_\infty \cup L_2$, if there exist real symmetric matrices $R > 0$, $Q_r > 0$, $S_r > 0$, $M_r = \begin{bmatrix} M_{r,11} & M_{r,12} \\ * & M_{r,22} \end{bmatrix} \geq 0$, real matrices X_r and Y_r, and positive constants μ_1 and μ_2, such that LMIs (3.10) and (3.20) are feasible*

$$\tilde{\Xi} = \begin{bmatrix} \bar{\Xi}_{11} & \tilde{\Xi}_{12} & \bar{\Xi}_{13} & \tilde{\Xi}_{14} & 0 \\ * & \bar{\Xi}_{22} & \tilde{\Xi}_{23} & 0 & \tilde{\Xi}_{25}^T \\ * & * & \bar{\Xi}_{33} & \tilde{\Xi}_{34} & 0 \\ * & * & * & -I & 0 \\ * & * & * & * & -I \end{bmatrix} < 0, \tag{3.20}$$

where

$$\bar{\Xi}_{11} = R(I_{N-1}\otimes\bar{A})+(I_{N-1}\otimes\bar{A})^T R+\mu_1 R+\sum_{r=1}^{k}(X_r + X_r^T)+\sum_{r=1}^{k} Q_r+\sum_{r=1}^{k}\bar{\tau}_r M_{r,11},$$

$$\bar{\Xi}_{22} = \text{blockdiag}\Big\{(d_1 - 1)Q_1 - Y_1 - Y_1^T + \bar{\tau}_1 M_{1,22}, \ldots, (d_k - 1)Q_k - Y_k - Y_k^T + \bar{\tau}_k M_{k,22}\Big\},$$

$$\bar{\Xi}_{13} = \Big[\bar{\tau}_1(I_{N-1}\otimes\bar{A})^T S_1(1+\mu_2), \ldots, \bar{\tau}_k(I_{N-1}\otimes\bar{A})^T S_k(1+\mu_2)\Big],$$

$$\tilde{\Xi}_{23} = \begin{bmatrix} -\bar{\tau}_1\left(\tilde{U}L_1\bar{U}\otimes\bar{B}\right)^T S_1 & \ldots & -\bar{\tau}_k\left(\tilde{U}L_1\bar{U}\otimes\bar{B}\right)^T S_k \\ \vdots & \ddots & \vdots \\ -\bar{\tau}_1\left(\tilde{U}L_k\bar{U}\otimes\bar{B}\right)^T S_1 & \ldots & -\bar{\tau}_k\left(\tilde{U}L_k\bar{U}\otimes\bar{B}\right)^T S_k \end{bmatrix}(1+\mu_2),$$

$$\bar{\Xi}_{33} = \text{blockdiag}\{-\bar{\tau}_1 S_1(1+\mu_2), \ldots, -\bar{\tau}_k S_k(1+\mu_2)\},$$

$$\tilde{\Xi}_{14} = -R\left(\tilde{U}\mathscr{D}\otimes\bar{B}\right),$$

$$\bar{\Xi}_{34} = \begin{bmatrix} -\tau_1 S_1^T (1+\mu_2)\left(\tilde{U}\mathscr{D} \otimes \bar{B}\right) \\ \vdots \\ -\tau_k S_k^T (1+\mu_2)\left(\tilde{U}\mathscr{D} \otimes \bar{B}\right) \end{bmatrix},$$

$$\tilde{\Xi}_{25} = \left[\mathscr{E}_1 \bar{U} \otimes I, \ldots, \mathscr{E}_k \bar{U} \otimes I\right].$$

Proof Considering the same Lyapunov-Krasovskii functional candidate as the one in the proof of Theorem 3.1, it can be obtained that

$$\begin{aligned}\dot{V}_1(t) \le\ & \varsigma_p^H(t)(R(I_{N-1}\otimes A) + (I_{N-1}\otimes A)^T R)\varsigma_p(t) \\ & - \sum_{r=1}^{k} 2\varsigma_p^H(t) R\left(\tilde{U}L_r\bar{U} \otimes \bar{B}\right)\varsigma_p(t-\tau_r(t)) \\ & + \mu_1 \varsigma_p^H(t) R \varsigma_p(t) + \frac{1}{\mu_1}\omega^T(t)(\tilde{U}\otimes\bar{B}_\omega)^T R(\tilde{U}\otimes\bar{B}_\omega)\omega(t),\end{aligned}$$

$$\dot{V}_2(t) \le \sum_{r=1}^{k} \varsigma_p^H(t) Q_r \varsigma_p(t) - \sum_{r=1}^{k}(1-d_r)\varsigma_p^H(t-\tau_r(t)) Q_r \varsigma_p(t-\tau_r(t)),$$

$$\begin{aligned}\dot{V}_3(t) \le & \sum_{r=1}^{k} \bar{\tau}_r(1+\mu_2)\left[\varsigma_p^H(t)\left(I_{N-1}\otimes\bar{A}\right)^T + \varsigma_p^H(t-\tau_1(t))\left(-\tilde{U}L_1\bar{U}\otimes\bar{B}\right)^T\right. \\ & \left. + \cdots + \varsigma_p^H(t-\tau_k(t))\left(-\tilde{U}L_k\bar{U}\otimes\bar{B}\right)^T\right] \\ & \times S_r\left[\left(I_{N-1}\otimes\bar{A}\right)\varsigma_p(t) + \left(-\tilde{U}L_1\bar{U}\otimes\bar{B}\right)\varsigma_p(t-\tau_1(t))\right. \\ & \left. + \cdots + \left(-\tilde{U}L_k\bar{U}\otimes\bar{B}\right)\varsigma_p(t-\tau_k(t))\right] \\ & + \sum_{r=1}^{k}\bar{\tau}_r\frac{1}{\mu_2}\omega^T(t)(\tilde{U}\otimes\bar{B}_\omega)^T S_r(\tilde{U}\otimes\bar{B}_\omega)\omega(t) \\ & + \sum_{r=1}^{k}\bar{\tau}_r\omega^T(t)(\tilde{U}\otimes\bar{B}_\omega)^T S_r(\tilde{U}\otimes\bar{B}_\omega)\omega(t) \\ & - \sum_{r=1}^{k}\int_{t-\tau_r(t)}^{t}\dot{\varsigma}_p^H(\alpha) S_r \dot{\varsigma}_p(\alpha) d\alpha.\end{aligned}$$

With a similar analysis described in the proof of Theorem 3.1, one gets

$$\begin{aligned}\dot{V}(t) &\le \dot{V}_1(t) + \dot{V}_2(t) + \dot{V}_3(t) + \Gamma_1 + \Gamma_2 \\ &\le \xi_0^H(t)\bar{\Xi}_0\xi_0(t) - \sum_{r=1}^{k}\textstyle\int_{t-\tau_r(t)}^{t}\xi_r^H(t,\alpha)\Upsilon_r\xi_r(t,\alpha)d\alpha + \omega^T(t)\Phi\omega(t),\end{aligned} \tag{3.21}$$

where

$$\bar{\Xi}_0 = \begin{bmatrix} \bar{\Xi}_{11} & \bar{\Xi}_{12} \\ * & \bar{\Xi}_{22} \end{bmatrix} + \sum_{r=1}^{k} \bar{\tau}_r(1+\mu_2)\left[I_{N-1}\otimes\bar{A}, -\tilde{U}L_1\bar{U}\otimes\bar{B}, \ldots, -\tilde{U}L_k\bar{U}\otimes\bar{B}\right]^T$$
$$\times S_r\left[I_{N-1}\otimes\bar{A}, -\tilde{U}L_1\bar{U}\otimes\bar{B}, \ldots, -\tilde{U}L_k\bar{U}\otimes\bar{B}\right],$$

$$\bar{\Xi}_{12} = \left[-R\left(\tilde{U}(L_1+\Delta L_1)\bar{U}\otimes\bar{B}\right) - X_1 + Y_1^T + \bar{\tau}_1 M_{1,12}, \ldots,\right.$$
$$\left. -R\left(\tilde{U}(L_k+\Delta L_k)\bar{U}\otimes\bar{B}\right) - X_k + Y_k^T + \bar{\tau}_k M_{k,12}\right],$$

$$\Phi = \frac{1}{\mu_1}(\tilde{U}\otimes\bar{B}_\omega)^T R(\tilde{U}\otimes\bar{B}_\omega) + \sum_{r=1}^{k}\bar{\tau}_r\frac{1}{\mu_2}(\tilde{U}\otimes\bar{B}_\omega)^T S_r(\tilde{U}\otimes\bar{B}_\omega)$$
$$+\sum_{r=1}^{k}\bar{\tau}_r(\tilde{U}\otimes\bar{B}_\omega)^T S_r(\tilde{U}\otimes\bar{B}_\omega).$$

By Schur complement, $\bar{\Xi}_0 < 0$ is equivalent to

$$\bar{\Xi} = \begin{bmatrix} \bar{\Xi}_{11} & \bar{\Xi}_{12} & \bar{\Xi}_{13} \\ * & \bar{\Xi}_{22} & \bar{\Xi}_{23} \\ * & * & \bar{\Xi}_{33} \end{bmatrix} < 0, \tag{3.22}$$

where

$$\bar{\Xi}_{23} = \begin{bmatrix} -\bar{\tau}_1\left(\tilde{U}\tilde{L}_1\bar{U}\otimes\bar{B}\right)^T S_1 & \ldots & -\bar{\tau}_k\left(\tilde{U}\tilde{L}_1\bar{U}\otimes\bar{B}\right)^T S_k \\ \vdots & \ddots & \vdots \\ -\bar{\tau}_1\left(\tilde{U}\tilde{L}_k\bar{U}\otimes\bar{B}\right)^T S_1 & \ldots & -\bar{\tau}_k\left(\tilde{U}\tilde{L}_k\bar{U}\otimes\bar{B}\right)^T S_k \end{bmatrix},$$
$$\tilde{L}_r = (L_r+\Delta L_r)(1+\mu_2)\ \ (r=1,2,\ldots,k).$$

By Lemma 3.5, replacing ΔL_r with $\mathscr{D}\bar{\Lambda}(t)\bar{\mathscr{E}}_r$, one can obtain that $\bar{\Xi} < 0$ is equivalent to

$$\begin{bmatrix} \bar{\Xi}_{11} & \tilde{\Xi}_{12} & \bar{\Xi}_{13} \\ * & \bar{\Xi}_{22} & \tilde{\Xi}_{23} \\ * & * & \bar{\Xi}_{33} \end{bmatrix} + H\left(\bar{\Lambda}(t)\otimes I\right)K + K^T\left(\bar{\Lambda}(t)\otimes I\right)^T H^T < 0, \tag{3.23}$$

where

$$H = \left[\tilde{\Xi}_{14}^T, 0, \tilde{\Xi}_{34}^T\right]^T,\ K = \left[0, \tilde{\Xi}_{25}, 0\right].$$

It follows from Lemma 3.6 that inequality (3.23) can be transformed into

$$\begin{bmatrix} \bar{\Xi}_{11} & \tilde{\Xi}_{12} & \bar{\Xi}_{13} \\ * & \bar{\Xi}_{22} & \tilde{\Xi}_{23} \\ * & * & \bar{\Xi}_{33} \end{bmatrix} + \gamma H H^T + \gamma^{-1} K^T K < 0. \tag{3.24}$$

Using Schur complement and replacing γR, γQ_r, γS_r, γM_r, γX_r and γY_r with R, Q_r, S_r, M_r, X_r and Y_r $(r = 1, 2, \ldots, k)$ respectively, one gets that inequality (3.24) is equivalent to LMI (3.20).

From (3.21), one knows that if $\omega(t) \equiv 0$, then $\bar{\Xi}_0 < 0$ and $\Upsilon_r \geq 0$ $(r = 1, 2, \ldots, k)$ are sufficient conditions for swarm system (3.3) to achieve state consensus.

For the case where $\omega(t) \in L_\infty$, if $\bar{\Xi}_0 < 0$ and $\Upsilon_r \geqslant 0$ $(r = 1, 2, \ldots, k)$, then from (3.21), one has

$$\varsigma_p^H(t)\varsigma_p(t) \leq \xi_0^H(t)\xi_0(t) \leq \alpha, \;\; t \geq T_0, \tag{3.25}$$

where T_0 is a sufficient large positive number and

$$\alpha = \frac{\lambda_{\Phi_{\max}}}{-\lambda_{\bar{\Xi}_{0\max}}} \left\| \omega^T \omega \right\|_\infty,$$

with $\lambda_{\Phi_{\max}}$ and $\lambda_{\bar{\Xi}_{0\max}}$ being the maximum eigenvalues of Φ and $\bar{\Xi}_0$ respectively. If $\omega(t) \in L_2$ and $\Upsilon_r \geqslant 0$ $(r = 1, 2, \ldots, k)$, from (3.21), it holds that for any $t \geqslant 0$

$$\int_0^t \dot{V}(\varsigma_p(\tau))d\tau - \int_0^t \xi_0^H(\tau)\bar{\Xi}_0\xi_0(\tau)d\tau \leqslant \int_0^t \omega^T(\tau)\Phi\omega(\tau)d\tau. \tag{3.26}$$

In this case, if $\bar{\Xi}_0 < 0$, then

$$\varsigma_p^H(t)\varsigma_p(t) \leqslant \beta, \tag{3.27}$$

where

$$\beta = \frac{1}{\lambda_{R_{\min}}} \left(V(\varsigma_p(0)) + \lambda_{\Phi_{\max}} \|\omega\|_2^2 \right),$$

with $\lambda_{R_{\min}}$ being the minimum eigenvalue of R. From (3.8) and (3.9), one has

$$\theta_i(t) - \tilde{\theta}_1(t) = (e_i^T \otimes I_{n+\bar{n}})(U \otimes I_{n+\bar{n}}) \left[0, \varsigma_p^H(t)\right]^H \;\; (i = 1, 2, \ldots, N).$$

From (3.6) and Lemma 3.3, it follows that

$$\begin{aligned}\lim_{t\to\infty}\|\theta_i(t)-c(t)\|_2 &\le \lim_{t\to\infty}\left\|\theta_i(t)-\tilde{\theta}_1(t)\right\|_2+\lim_{t\to\infty}\left\|\tilde{\theta}_1(t)-c(t)\right\|_2\\ &\le \left\|(e_i^T\otimes I_{n+\bar{n}})(U\otimes I_{n+\bar{n}})\right\|_\infty \lim_{t\to\infty}\left\|\varsigma_p(t)\right\|_2\\ &\le \max_i\|\tilde{\upsilon}_i\|_2\lim_{t\to\infty}\left\|\varsigma_p(t)\right\|_2,\end{aligned}$$

where $\tilde{\upsilon}_i$ $(i=1,2,\ldots,N)$ are rows of U. Therefore, if $\omega(t)\in L_\infty\cup L_2$, $\bar{\Xi}_0<0$ and $\Upsilon_r\geqslant 0$ $(r=1,2,\ldots,k)$, swarm system (3.3) achieves state practical consensus and satisfies

$$\lim_{t\to\infty}\|\theta_i(t)-c(t)\|_2\leqslant\delta,$$

where $c(t)$ is a state practical consensus function satisfies Lemma 3.3 and

$$\delta=\begin{cases}\max\limits_i\|\tilde{\upsilon}_i\|_2\sqrt{\alpha}, & \forall\,\omega(t)\in L_\infty,\\ \max\limits_i\|\tilde{\upsilon}_i\|_2\sqrt{\beta}, & \forall\,\omega(t)\in L_2.\end{cases}$$

The proof for Theorem 3.2 is completed.

Remark 3.3 In [8], the consensus function is the average of initial states of all agents and the disturbances belong to L_2. However, in [8], time-varying delays were not considered and it was required that the topologies are undirected. In the current work, sufficient conditions for general high-order LTI swarm systems with external disturbances, interaction uncertainties, and time-varying delays to achieve practical consensus are obtained, where the external disturbances can belong to L_2 or L_∞ and the interaction topologies are described by general directed graphs. It should be mentioned that if the interaction topology does not have a spanning tree or even if all agents are isolated, although this case is trivial for swarm systems, our results can still ensure that swarm system (3.3) achieves state practical consensus. In this case, by Lemma 2.2, one knows that the Laplacian matrix at least has two zero eigenvalues. It can be shown that there exists at least one subsystem whose stability is determined by $\bar{A}$ in subsystem (3.7). To guarantee the stability of subsystem (3.7), $\bar{A}$ must be Hurwitz. In this case, the state practical consensus is certainly achieved.

3.5 Numerical Simulations

In this section, a numerical example is given to demonstrate theoretical results obtained in the previous sections.

Consider a third-order swarm system with six agents. The dynamics of each agent are described by (3.3) with $\theta_i(t)=\left[x_{i1}^T(t),x_{i2}^T(t),x_{i3}^T(t),z_i^T(t)\right]^T$ $(i=1,2,\ldots,6)$ and

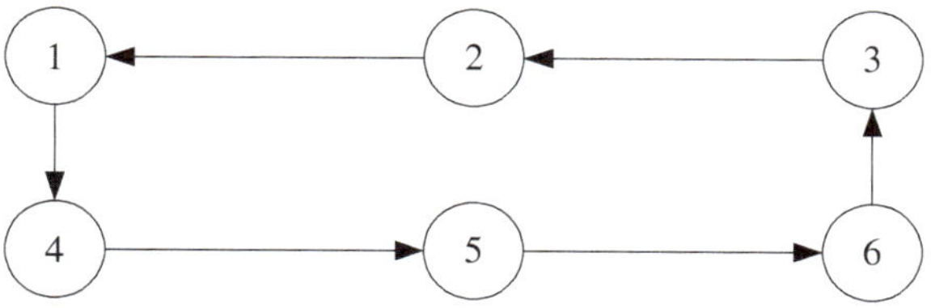

Fig. 3.1 Directed interaction topology G

$$\bar{A} = \begin{bmatrix} 0 & 1 & 0 & 0 \\ 0 & 0 & 1 & 0 \\ 0.5 & -6 & -5.76 & 6 \\ 0 & 0 & 0 & -5 \end{bmatrix}, \ \bar{B} = \begin{bmatrix} 0 & 0 & 0 & 0 \\ 0 & 0 & 0 & 0 \\ -6 & -0.2 & 0 & 0 \\ 9.2 & 6 & 0 & 0 \end{bmatrix}, \ \bar{B}_\omega = \begin{bmatrix} 0 \\ 0 \\ 1 \\ 0 \end{bmatrix}.$$

The interaction topology of the swarm system is shown in Fig. 3.1. For simplicity, it is assumed that the interaction topology G in this chapter and the following chapters has 0-1 weights. Let the initial states be $x_{i1}(0) = 5(\Theta - 0.5)$, $x_{i2}(0) = 4(\Theta - 0.5)$, $x_{i3}(0) = 3(\Theta - 0.5)$ and $z_i(0) = 2(\Theta - 0.5)$ where $i = 1, 2, \ldots, 6$, and Θ is a pseudorandom value from a uniform distribution on the unit interval.

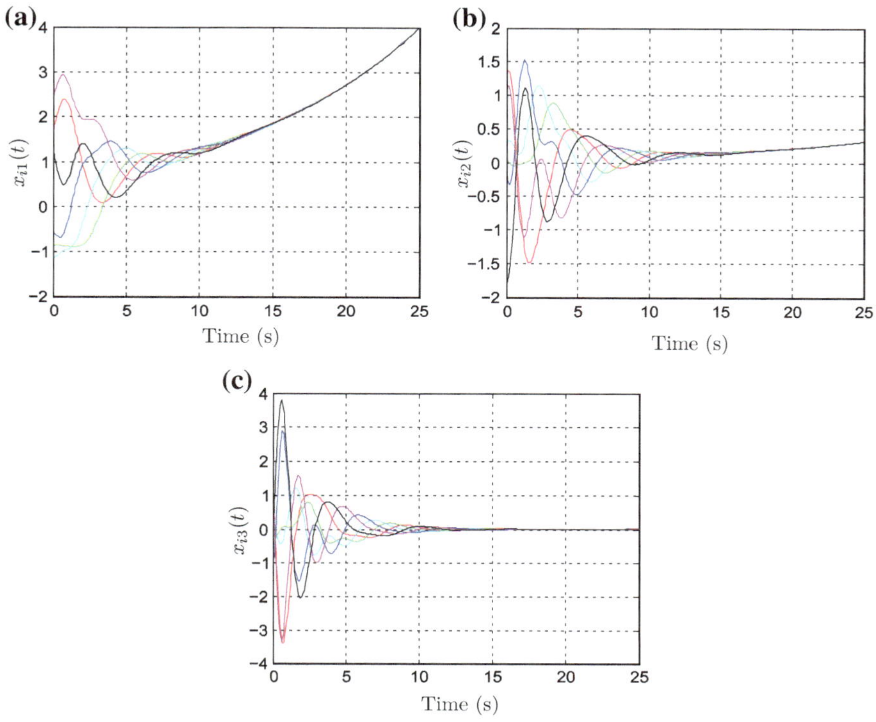

Fig. 3.2 Trajectories of $x_{i1}(t)$, $x_{i2}(t)$ and $x_{i3}(t)$ ($i = 1, 2, \ldots, 6$). **a** Trajectories of $x_{i1}(t)$, **b** Trajectories of $x_{i2}(t)$ and **c** Trajectories of $x_{i3}(t)$

Case 1: For the swarm system with $\omega(t) \equiv 0$ and $\Delta L_r \equiv 0$, assume that the nonuniform time-varying delays satisfy $\tau_{12}(t) = 0.04 + 0.031\sin(t)$, $\tau_{23}(t) = 0.04 + 0.032\sin(t)$, $\tau_{34}(t) = 0.04 + 0.033\sin(t)$, $\tau_{45}(t) = 0.04 + 0.034\sin(t)$, $\tau_{56}(t) = 0.04 + 0.035\sin(t)$, and $\tau_{61}(t) = 0.04 + 0.036\sin(t)$. By using the Feasp solver in Matlab's LMI toolbox [15], it can be verified that LMIs (3.10) and (3.11) in Theorem 3.1 are feasible. Figure 3.2 shows that all states of the six agents achieve consensus.

Case 2: Set the nonuniform time-varying delays as $\tau_{12}(t) = 0.03 + 0.021\sin(t)$, $\tau_{23}(t) = 0.03 + 0.022\sin(t)$, $\tau_{34}(t) = 0.03 + 0.023\sin(t)$, $\tau_{45}(t) = 0.03 + 0.024\sin(t)$, $\tau_{56}(t) = 0.03 + 0.025\sin(t)$, $\tau_{61}(t) = 0.03 + 0.026\sin(t)$, and set $\bar{\Lambda}(t) = 0.1\sin(t)I_6$, $\omega_i(t) = 0.05i\sin(t)$ $(i = 1, 2, \ldots, 6)$, $\mu_1 = 0.08$, and $\mu_2 = 1.8$. One can check that LMIs (3.11) and (3.20) in Theorem 3.2 are feasible. In Fig. 3.3, circles denote the present coordinate of each agent's states, and arrows represent the moving direction of each agent's states. The length of the arrow denotes the relative magnitude of the states. From the snapshots of the states' trajectories in Fig. 3.3, one can see that the moving directions and magnitudes of six agents' states are reaching an agreement with small error; that is, the swarm system achieves state practical consensus.

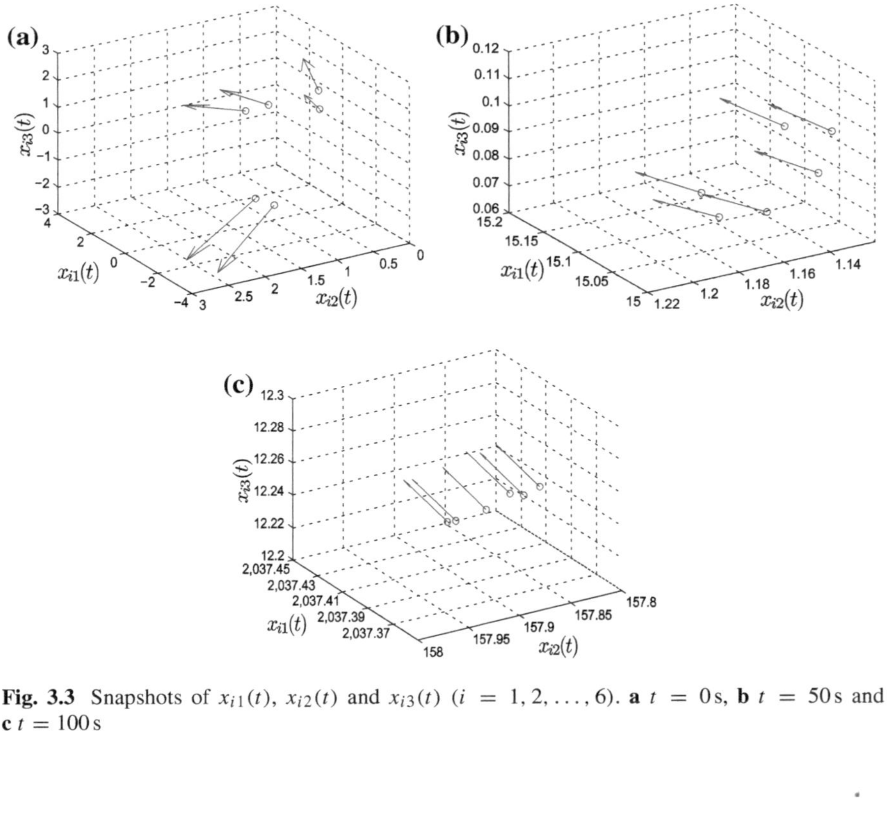

Fig. 3.3 Snapshots of $x_{i1}(t)$, $x_{i2}(t)$ and $x_{i3}(t)$ $(i = 1, 2, \ldots, 6)$. **a** $t = 0\,\mathrm{s}$, **b** $t = 50\,\mathrm{s}$ and **c** $t = 100\,\mathrm{s}$

3.6 Conclusions

State practical consensus problems for general high-order LTI swarm systems with time-varying external disturbances and interaction uncertainties in presence of nonuniform time-varying delays were studied in this chapter. A protocol was constructed based on the dynamic output feedback. Using state space decomposition techniques, state practical consensus problems of the original system were converted into stability problems of the disagreement subsystems. Based on the Lyapunov-Krasovskii functional approach, sufficient conditions for swarm systems to achieve state practical consensus were proposed in terms of LMIs, where the time-varying external disturbance can be in L_2 or L_∞. Explicit expressions of the practical consensus function and consensus error bounds were derived. The results in this chapter are mainly based on [16].

References

1. Ren W, Moore KL, Chen YQ (2007) High-order and model reference consensus algorithms in cooperative control of multivehicle systems. J Dyn Syst Meas Control 129(5):678–688
2. Wang JH, Cheng DZ (2007) Consensus of multi-agent systems with higher order dynamics. In: Proceeding of the 26th Chinese control conference, pp 761–765
3. Jiang FC, Wang L, Xie GM (2010) Consensus of high-order dynamic multi-agent systems with switching topology and time-varying delays. J Control Theor Appl 8(1):52–60
4. Zhang W, Zeng D, Qu S (2010) Dynamic feedback consensus control of a class of high-order multi-agent systems. IET Control Theory Appl 4(10):2219–2222
5. Xiao F, Wang L (2007) Consensus problems for high-dimensional multi-agent systems. IET Control Theory Appl 1(3):830–837
6. Seo JH, Shim H, Back J (2009) Consensus of high-order linear systems using dynamic output feedback compensator: low gain approach. Automatica 45(11):2659–2664
7. Xi JX, Cai N, Zhong YS (2010) Consensus problems for high-order linear time-invariant swarm systems. Phys A 389(24):5619–5627
8. Liu Y, Jia YM (2012) H_∞ consensus control for multi-agent systems with linear coupling dynamics and communication delays. Int J Syst Sci 43(1):50–62
9. Lin P, Jia YM (2010) Robust H_∞ consensus analysis of a class of second-order multi-agent systems with uncertainty. IET Control Theory Appl 4(3):487–498
10. Ni W, Cheng DZ (2010) Leader-following consensus of multi-agent systems under fixed and switching topologies. Syst Control Lett 59(3–4):209–217
11. Vanderveen MC, Van der Veen AJ, Paulraj A (1998) Estimation of multipath parameters in wireless communications. IEEE Trans Signal Process 46(3):682–690
12. Olfati-Saber R, Murray RM (2004) Consensus problems in networks of agents with switching topology and time-delays. IEEE Trans Autom Control 49(9):1520–1533
13. Gahinet P, Apkarian A (1994) A linear matrix inequality approach to H_∞ control. Int J Robust Nonlinear Control 4(4):421–448
14. Petersen IR (1987) A stabilization algorithm for a class of uncertain linear systems. Syst Control Lett 8(4):351–357
15. Gahinet P, Nemirovski A, Laub AJ et al (1995) LMI control toolbox for use with Matlab. MathWorks, Natick
16. Dong XW, Xi JX, Shi ZY et al (2013) Practical consensus for high-order linear time-invariant swarm systems with interaction uncertainties, time-varying delays and external disturbances. Int J Syst Sci 44(10):1843–1856

Chapter 4
Formation Control of Swarm Systems

Abstract This chapter investigates time-varying formation control problems for high-order linear time-invariant (LTI) swarm systems and unmanned aerial vehicle (UAV) swarm systems. First, time-varying state formation control problems for high-order LTI swarm systems with time delays and time-varying output formation control problems for high-order LTI swarm systems are studied, respectively. Necessary and sufficient conditions for swarm systems to achieve time-varying state/output formation, necessary and sufficient conditions for state/output formation feasibilities, and explicit expressions of state/output formation reference functions are presented. Then approaches to specify the motion modes of state/output formation references, approaches to expand feasible state/output formation sets, and approaches to design the state/output formation protocols are proposed, respectively. Moreover, necessary and sufficient conditions for UAV swarm systems to achieve time-varying formations and approaches to design the protocol are proposed. Autonomous time-varying formation control experiments are performed using five quadrotor UAVs in outdoor environment to demonstrate the theoretical results.

4.1 Introduction

Formation control of swarm systems has received considerable interests in recent years due to its broad potential applications in civilian and military areas such as drag reduction [1], surveillance and reconnaissance [2, 3], radiation detection and contour mapping [4], target search and localization [5], telecommunication relay [6], etc. To achieve formation control, many approaches have been proposed, such as leader–follower [7], behavior [8] and virtual structure [9]-based approaches, etc. However, as pointed out by Beard et al. [10], leader–follower, behavior and virtual structure-based formation control approaches have their own strengths and weaknesses. For example, leader–follower approaches are easy to implement but have no explicit feedback on formations and lack of robustness due to the existence of the leader, and behavior approaches can ensure collision avoidance but are difficult to analyze mathematically, etc.

X. Dong, *Formation and Containment Control for High-order Linear Swarm Systems*, Springer Theses, DOI 10.1007/978-3-662-47836-3_4

With the development of consensus theory, more and more researchers realize that consensus approaches can be used to deal with formation control problems. Ren [11] presented consensus-based formation control approaches for second-order swarm systems and showed that leader–follower, behavior and virtual structure-based formation control approaches can be treated as special cases of consensus-based ones. Moreover, in consensus-based formation control approaches, there exist formation feedback and no explicit leader, and only local neighbor-to-neighbor information between the agents is required to achieve formations, which overcomes the weakness of the previous approaches. The authors in [11–15] studied formation control problems for low-order swarm systems using consensus-based strategies. In [16], a special high-order LTI formation model, which can be regarded as a series of second-order models, was discussed based on consensus approaches. However, the results in [11–16] cannot be used to solve the formation control problems for general high-order LTI swarm system. Although Fax and Murray [17] and Porfiri et al. [18] discussed formation stability problems for general high-order LTI swarm systems, they did not deal with the problem that how to achieve desired formations. Moreover, none of the results mentioned above consider the formation feasibility problems. Lin et al. [19] investigated formation feasibility problems for low-order swarm systems. Ma and Zhang [20] proposed a criterion for formation feasibility of high-order LTI swarm systems. It should be pointed out that in [19] and [20], the formation is time-invariant and the feasible formation set is very limited. Time-varying formation control problems for high-order LTI swarm systems are still open.

Time delays often exist in the cooperative control of swarm systems. Therefore, it is necessary and meaningful to consider time delays in time-varying formation control problems of high-order LTI swarm systems. However, no one has discussed the time-varying formations and formation feasibility problems for high-order LTI swarm systems with time delays. Moreover, in some practical applications, only the output information of each agent is available and only outputs are required to achieve formations. In these cases, time-varying output formation problems and output formation feasibility problems arise. Up to now, neither time-varying output formation problems nor output formation feasibility problems for high-order LTI swarm systems have been studied. It is well known that UAVs have broad potential applications in various fields due to their high maneuverability and excellent economical efficiency. The research on control and applications of UAVs are becoming hot issues among which autonomous time-varying formation control for UAV swarm systems is one of the most complicated and challenging problems.

This chapter mainly investigates time-varying state formation control and formation feasibility problems for high-order LTI swarm systems with time delays, time-varying output formation control and formation feasibility problems for high-order LTI swarm systems, and time-varying formation control and experimental realization problems for UAV swarm systems. The main results are summarized as follows.

For high-order LTI swarm systems to achieve time-varying state formations, a general formation protocol with time-delays is proposed first. Many previous consensus-based formation protocols can be regarded as special cases of the protocol in this chapter. Using the state transformation and state decomposition approaches, time-varying state formation control problems are converted into asymptotic stability problems. Necessary and sufficient conditions for high-order LTI swarm systems with time-delays to achieve time-varying state formation and for state formation feasibility, and an explicit expression of state formation reference function are presented. Approaches to specify the motion modes of the state formation reference, approaches to expand the feasible state formation set, and approaches to design the state formation protocols are proposed. All the results can be applied to deal with the time-varying formation problems for high-order LTI swarm systems without time-delays. Moreover, it is shown that many existing consensus-based formation control results are special cases of the results in this chapter. Comparative examples are also given to show that using the approach in [18], formations cannot be achieved while using the approach in the current chapter, the formation can be achieved and the motion modes of the formation reference can be specified.

Based on the static output feedback approach, an output formation protocol is constructed for high-order LTI swarm systems to achieve time-varying output formations. Using output transformation approach, output space decomposition approach, and partial stability theory, necessary and sufficient conditions for swarm systems to achieve time-varying output formations, and necessary and sufficient conditions for output formation feasibility are proposed, respectively. An explicit expression of the time-varying output formation reference function and approaches to specify the motion modes of the output formation reference are given. Moreover, approaches to expand the feasible output formation set and approaches to design the protocols are presented. It is shown that the results for time-varying output formation can be applied to deal with state/output consensus problems and time-varying state formation control problems.

Theoretical results for high-order LTI swarm systems are applied to solve the time-varying formation control problems for UAV swarm systems. Necessary and sufficient conditions for UAV swarm systems to achieve time-varying formations are proposed. An approach to determine the gain matrices in the protocol is given.It is proven that by choosing appropriate initial states for each UAV, the collision among UAVs can be avoided. Autonomous time-varying formation experiments using five quadrotor UAVs are performed in outdoor environment to demonstrate the theoretical results.

The rest of this chapter is organized as follows. In Sect. 4.2, time-varying state formation control problems for high-order LTI swarm systems with time delays are studied. In Sect. 4.3, time-varying output formation problems for high-order LTI swarm systems are investigated. Time-varying formation control problems for UAV swarm systems are addressed in Sect. 4.4. Section 4.5 concludes the whole work of this chapter.

4.2 Time-Varying State Formation Control with Time Delays

In this section, time-varying state formation control problems for high-order LTI swarm systems with time delays are investigated using a consensus-based approach. Necessary and sufficient conditions for swarm systems to achieve time-varying state formation, and for state formation feasibilities are proposed. Approaches to specify the motion modes of state formation references, and to expand feasible state formation sets, as well as to design the state formation protocols are presented, respectively. Numerical examples are given to demonstrate the theoretical results. Comparative examples are also given to show that using the approach in [18], formations cannot be achieved while using the approach in the current section, not only the formation can be achieved but also the motion modes of the formation reference can be specified.

4.2.1 Problem Description

Consider a swarm system with N agents on a directed interaction topology G. It is assumed that G has a spanning tree. Suppose that each agent has the LTI dynamics described by

$$\dot{x}_i(t) = Ax_i(t) + Bu_i(t), \tag{4.1}$$

where $i \in \{1, 2, \ldots, N\}$, $x_i(t) \in \mathbb{R}^n$ is the state, $u_i(t) \in \mathbb{R}^m$ is the control input and it is assumed that $\text{rank}(B) = m$. A time-varying state formation is specified by a vector $h(t) = [h_1^T(t), h_2^T(t), \ldots, h_N^T(t)]^T \in \mathbb{R}^{nN}$.

Definition 4.1 Swarm system (4.1) is said to *achieve time-varying state formation* $h(t)$ if for any given bounded initial states, there exists a vector-valued function $r(t) \in \mathbb{R}^n$ such that

$$\lim_{t\to\infty} (x_i(t) - h_i(t) - r(t)) = 0 \ (i = 1, 2, \ldots, N), \tag{4.2}$$

where $r(t)$ is called a *state formation reference function*.

Remark 4.1 Definition 4.1 presents a general definition for a formation specified by vectors $h_i(t)$ $(i = 1, 2, \ldots, N)$. The definitions adopted in [11, 12, 14, 16, 20] can be regarded as special cases of Definition 4.1.

Figure 4.1 shows a triangle formation for a swarm system with three agents. From Fig. 4.1, one sees that if $x_i(t) - h_i(t) - r(t) \to 0$ as $t \to \infty$, for all $i = 1, 2, 3$, then the two triangles formed by $h_i(t)$ and $x_i(t)$ $(i = 1, 2, 3)$ are congruent; that is, the triangle formation is achieved.

Definition 4.2 If there exist control inputs $u_i(t)$ $(i = 1, 2, \ldots, N)$ such that for any given bounded initial states, swarm system (4.1) achieves the state formation $h(t)$, then state formation $h(t)$ is called *feasible* for swarm system (4.1).

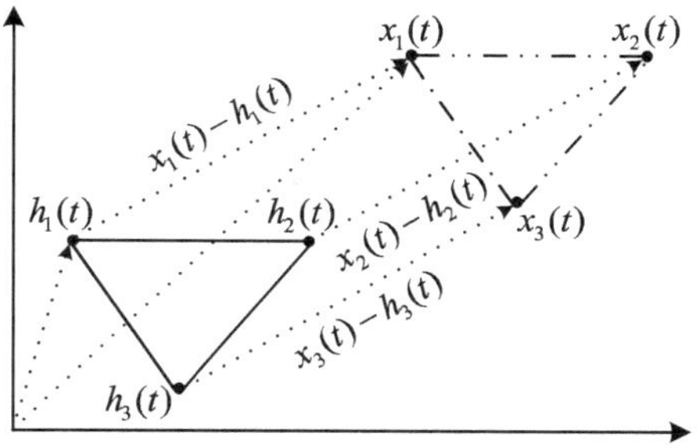

Fig. 4.1 A *triangle* formation for a swarm system with $N = 3$

Definition 4.3 Swarm system (4.1) is said to *achieve state consensus* if for any given bounded initial states, there exists a vector-valued function $c(t) \in \mathbb{R}^n$ such that

$$\lim_{t\to\infty} (x_i(t) - c(t)) = 0 \; (i = 1, 2, \ldots, N),$$

where $c(t)$ is called a *state consensus function.*

Remark 4.2 From Definitions 4.1 and 4.3, one can see that if $h(t) \equiv 0$, swarm system (4.1) achieves the point formation which is also known as consensus. In this case, the state formation reference function is equivalent to the state consensus function. Therefore, state consensus problem is just a special case of state formation control problem.

Consider a given time-varying state formation $h(t)$ with $h_i(t)$ $(i = 1, 2, \ldots, N)$ piecewise continuously differentiable and consider the following time-delayed formation protocol

$$\begin{aligned} u_i(t) &= K_1 x_i(t) + K_2 (x_i(t) - h_i(t)) \\ &\quad + K_3 \sum_{j\in N_i} w_{ij} \big((x_j(t-\tau) - h_j(t-\tau)) - (x_i(t-\tau) - h_i(t-\tau))\big) + v_i(t), \end{aligned} \tag{4.3}$$

where $i = 1, 2, \ldots, N$, N_i represent the neighboring set of agent i, K_1, K_2, and K_3 are constant gain matrices with appropriate dimensions, τ is a known constant time delay, and $v_i(t) \in \mathbb{R}^m$ is the time-varying state formation compensation signal to be determined later.

Remark 4.3 In protocol (4.3), the gain matrix K_1 and $v_i(t)$ $(i = 1, 2, \ldots, N)$ can be used to expand the set of feasible state formation $h(t)$. K_2 and K_3 can be used to design the motion modes of the state formation reference and propel the states of all agents to achieve the desired formation, respectively. It should be emphasized that K_1, K_2, and $v_i(t)$ $(i = 1, 2, \ldots, N)$ are not necessary for swarm systems to achieve some state formations. Moreover, protocol (4.3) presents a general framework for consensus-based formation control protocols. Many existing protocols, such as protocols shown in [11, 12, 14, 16, 20] can be regarded as special cases of protocol (4.3).

Definition 4.4 A time-varying state formation $h(t)$ is *feasible* for swarm system (4.1) under protocol (4.3), if there exist K_i $(i = 1, 2, 3)$ and $v_i(t)$ $(i = 1, 2, \ldots, N)$ such that it can be achieved for any given bounded initial states.

Let $x_i(t) = \varphi_i(t)$, $t \in [-\tau, 0]$ with $\varphi_i(t)$ being a continuous vector-valued function on $t \in [-\tau, 0]$. Let $x(t) = [x_1^T(t), x_2^T(t), \ldots, x_N^T(t)]^T$, $\varphi(t) = [\varphi_1^T(t), \varphi_2^T(t), \ldots, \varphi_N^T(t)]^T$ and $v(t) = [v_1^T(t), v_2^T(t), \ldots, v_N^T(t)]^T$. Under protocol (4.3), swarm system (4.1) can be written in a compact form as follows:

$$\begin{cases} \dot{x}(t) = (I_N \otimes (A + BK_1 + BK_2))\, x(t) + (I_N \otimes B)\, v(t) - (L \otimes BK_3)\, x(t - \tau) \\ \qquad - (I_N \otimes BK_2)\, h(t) + (L \otimes BK_3)\, h(t - \tau), t > 0, \\ x(t) = \varphi(t), t \in [-\tau, 0]. \end{cases} \tag{4.4}$$

The current section mainly focuses on the following three problems for swarm system (4.4): (i) under what conditions the time-varying state formation $h(t)$ can be achieved; (ii) under what conditions a time-varying state formation $h(t)$ is feasible; and (iii) how to design protocol (4.3) to achieve the time-varying state formation $h(t)$.

4.2.2 Time-Varying State Formation Analysis

In this subsection, first, a consensus-based approach is applied to deal with the time-varying state formation problems for swarm system (4.4). Then necessary and sufficient conditions for swarm system (4.4) to achieve the time-varying state formation $h(t)$ are presented, and an explicit expression of the state formation reference function is given.

Let λ_i $(i = 1, 2, \ldots, N)$ be the eigenvalues of the Laplacian matrix L corresponding to the directed interaction topology G, where $\lambda_1 = 0$ with the associated eigenvector $\bar{u}_1 = \mathbf{1}$ and $0 < \mathrm{Re}(\lambda_2) \leq \cdots \leq \mathrm{Re}(\lambda_N)$. Let $U^{-1}LU = J$, where $U = [\bar{u}_1, \bar{u}_2, \ldots, \bar{u}_N]$, $U^{-1} = [\tilde{u}_1^H, \tilde{u}_2^H, \ldots, \tilde{u}_N^H]^H$ and J is the Jordan canonical form of L. Let $\tilde{x}_i(t) = x_i(t) - h_i(t)$ $(i = 1, 2, \ldots, N)$ and $\tilde{x}(t) = [\tilde{x}_1^T(t), \tilde{x}_2^T(t), \ldots, \tilde{x}_N^T(t)]^T$. Then swarm system (4.4) can be rewritten as follows:

$$\begin{cases} \dot{\tilde{x}}(t) = (I_N \otimes (A + BK_1 + BK_2))\, \tilde{x}(t) - (L \otimes BK_3)\, \tilde{x}(t - \tau) \\ \qquad + (I_N \otimes (A + BK_1))\, h(t) - (I_N \otimes I)\, \dot{h}(t) \\ \qquad + (I_N \otimes B)\, v(t), t > 0, \\ \tilde{x}(t) = \varphi(t) - h(t), t \in [-\tau, 0]. \end{cases} \tag{4.5}$$

Based on Definitions 4.1 and 4.3, the following lemma holds directly:

Lemma 4.1 *Swarm system (4.4) achieves time-varying state formation $h(t)$ if and only if swarm system (4.5) achieves state consensus.*

Choose the ν in Sect. 2.2 as $\nu = n$. From the structure of $\mathbb{C}(U)$ and $\overline{\mathbb{C}}(U)$, one gets that the necessary and sufficient condition for swarm system (4.5) to achieve state consensus is $\lim_{t\to\infty}\tilde{x}(t) \in \mathbb{C}(U)$. Based on Lemma 2.1 and the structure of U, one can set $J = \operatorname{diag}\{0, \bar{J}\}$, where $\bar{J}$ consists of Jordan blocks corresponding to λ_i $(i = 2, \ldots, N)$. Let $\tilde{U} = [\tilde{u}_2, \tilde{u}_3, \ldots, \tilde{u}_N]^H$, $\theta_h(t) = (\tilde{u}_1^H \otimes I)\tilde{x}(t)$, and $\varsigma_h(t) = (\tilde{U} \otimes I)\tilde{x}(t)$, then swarm system (4.5) can be transformed into

$$\begin{aligned}\dot{\theta}_h(t) = {}& (A + BK_1 + BK_2)\,\theta_h(t) + (\tilde{u}_1^H \otimes B)v(t) + \left(\tilde{u}_1^H \otimes (A + BK_1)\right) h(t) \\ & - (\tilde{u}_1^H \otimes I)\dot{h}(t),\end{aligned} \tag{4.6}$$

$$\begin{aligned}\dot{\varsigma}_h(t) = {}& (I_{N-1} \otimes (A + BK_1 + BK_2))\,\varsigma_h(t) - (\bar{J} \otimes BK_3)\varsigma_h(t-\tau) \\ & + (\tilde{U} \otimes B)v(t) + \left(\tilde{U} \otimes (A + BK_1)\right) h(t) - (\tilde{U} \otimes I)\dot{h}(t).\end{aligned} \tag{4.7}$$

The following theorem presents a necessary and sufficient condition for swarm system (4.4) to achieve time-varying state formation $h(t)$.

Theorem 4.1 *For any given bounded initial states, swarm system (4.4) achieves time-varying state formation $h(t)$ if and only if* $\lim_{t\to\infty}\varsigma_h(t) = 0$.

Proof Let $\tilde{x}_C(t) = (U \otimes I_n)[\theta_h^H(t), 0]^H$ and $\tilde{x}_{\bar{C}}(t) = (U \otimes I_n)[0, \varsigma_h^H(t)]^H$. Considering the fact that $c_k \in \mathbb{R}^n$ $(k = 1, 2, \ldots, n)$ are linearly independent vectors, there exist $\alpha_k(t)$ $(k = 1, 2, \ldots, n)$ and $\alpha_{jn+k}(t)$ $(j = 1, 2, \ldots, N-1;\ k = 1, 2, \ldots, n)$ such that $\theta_h(t) = \sum_{k=1}^{n} \alpha_k(t)c_k$ and $\varsigma_h(t) = \left[\sum_{k=1}^{n} (\alpha_{n+k}(t)c_k)^H, \ldots, \sum_{k=1}^{n} (\alpha_{(N-1)n+k}(t)c_k)^H\right]^H$. Since $[\theta_h^H(t), 0]^H = e_1 \otimes \theta_h(t)$, the following holds

$$\tilde{x}_C(t) = (U \otimes I_n)\,(e_1 \otimes \theta_h(t)) = \sum_{k=1}^{n} \alpha_k(t)p_k \in \mathbb{C}(U). \tag{4.8}$$

By the structures of p_j $(j = n+1, n+2, \ldots, Nn)$, one has

$$\tilde{x}_{\bar{C}}(t) = \sum_{i=2}^{N}\sum_{k=1}^{n} \alpha_{(i-1)n+k}(t)\,(\bar{u}_i \otimes c_k) = \sum_{j=n+1}^{nN} \alpha_j(t)p_j \in \overline{\mathbb{C}}(U). \tag{4.9}$$

Owing to $[\theta_h^H(t), \varsigma_h^H(t)]^H = (U^{-1} \otimes I)\tilde{x}(t)$, one can obtain $\tilde{x}(t) = \tilde{x}_C(t) + \tilde{x}_{\bar{C}}(t)$. From Lemmas 4.1 and 2.3, swarm system (4.4) achieves time-varying state formation $h(t)$ if and only if $\lim_{t\to\infty}\tilde{x}_{\bar{C}}(t) = 0$, which means $\lim_{t\to\infty}\varsigma_h(t) = 0$. This completes the proof.

Remark 4.4 Ren [11] extended consensus results for second-order swarm systems to solve the formation control problems and showed that leader–follower, behavior and virtual structure-based formation approaches can be treated as special cases of consensus-based formation approaches. For the high-order LTI swarm system (4.1), choose

$$A = \begin{bmatrix} 0 & 1 \\ 0 & 0 \end{bmatrix}, \ B = \begin{bmatrix} 0 \\ 1 \end{bmatrix}.$$

For the time-varying state formation protocol (4.3), let

$$K_1 = 0, \ K_2 = \begin{bmatrix} 0 & -k_2 \end{bmatrix}, \ K_3 = \begin{bmatrix} 1 & k_3 \end{bmatrix}, \ v_i(t) = \dot{\gamma}(t), \ h_i(t) = \begin{bmatrix} 0 & \gamma(t) \end{bmatrix}^T, \ \tau = 0,$$

where $i = 1, 2, \ldots, N$, $\gamma(t)$ is the predefined continuously differentiable velocity, $k_2 > 0$ and $k_3 > 0$ are constant gains. Then the formation problems in this section become the consensus problems in [11]. It is evident that Corollary 3.1 in [11] is equivalent to Theorem 4.1 with the above-specified parameters.

Remark 4.5 For undirected interaction topologies, Xie and Wang [14] investigated formation problems for swarm system (4.4) with

$$\begin{aligned} &A = \begin{bmatrix} 0 & 1 \\ 0 & 0 \end{bmatrix}, \ B = \begin{bmatrix} 0 \\ 1 \end{bmatrix}, \ h_i = \begin{bmatrix} \delta_i \\ \gamma \end{bmatrix}, \ K_1 = 0, \ K_2 = \begin{bmatrix} 0 & k_2 \end{bmatrix}, \\ &K_3 = \begin{bmatrix} k_3 & 0 \end{bmatrix}, \ v_i = 0, \ \tau = 0, \end{aligned}$$

where $i = 1, 2, \ldots, N$, δ_i and γ are constant, k_2 and k_3 are constant gains, and presented a sufficient condition that if $k_2 < 0$, $k_3 > 0$ and the interaction topologies are connected then swarm system (4.4) achieves formation h. However, by Theorem 4.1, it can be verified that the condition in Xie and Wang [14] is not only sufficient but also necessary.

In the sequel, based on the above analysis, an explicit expression of the time-varying state formation reference function is given to describe the macroscopic movement of the whole state formation.

Theorem 4.2 *If swarm system (4.4) achieves time-varying state formation $h(t)$ with the state formation reference function $r(t)$, then*

$$\lim_{t \to \infty} (r(t) - r_0(t) - r_v(t) - r_h(t)) = 0,$$

where

$$r_0(t) = e^{(A+BK_1+BK_2)t} \left(\tilde{u}_1^H \otimes I \right) x(0),$$

$$r_v(t) = \int_0^t \left(e^{(A+BK_1+BK_2)(t-s)} (\tilde{u}_1^H \otimes B) v(s) \right) ds,$$

$$r_h(t) = -(\tilde{u}_1^H \otimes I) h(t) - \int_0^t e^{(A+BK_1+BK_2)(t-s)} \left(\tilde{u}_1^H \otimes BK_2 \right) h(s) ds.$$

Proof If swarm system (4.4) achieves time-varying state formation $h(t)$, then $\tilde{x}_i(t) - \theta_h(t) \to 0$ as $t \to \infty$. Therefore, subsystem (4.6) determines the state formation reference function. By Lemma 2.3, one has

$$\tilde{x}_C(0) = P_{\mathbb{C}(U),\overline{\mathbb{C}}(U)}\tilde{x}(0), \tag{4.10}$$

where $P_{\mathbb{C}(U),\overline{\mathbb{C}}(U)} = [p_1, \ldots, p_n, 0, \ldots, 0]P^{-1}$ is an oblique projector onto $\mathbb{C}(U)$ along $\bar{\mathbb{C}}(U)$ and $P = [p_1, p_2, \ldots, p_{Nn}]$. Let $\tilde{C} = [c_1, c_2, \ldots, c_n]$, then it can be obtained that $P = U \otimes \tilde{C}$. Owing to $[p_1, p_2, \ldots, p_n, 0, \ldots, 0] = [\mathbf{1}, 0] \otimes \tilde{C}$, one has

$$P_{\mathbb{C}(U),\overline{\mathbb{C}}(U)} = \left([\mathbf{1}, 0]\, U^{-1}\right) \otimes I. \tag{4.11}$$

From (4.8), one can obtain that

$$\theta_h(0) = [I, 0, \ldots, 0]\, \tilde{x}_C(0). \tag{4.12}$$

Because $[\mathbf{1}, 0]\, U^{-1} = \left(\mathbf{1} \otimes \tilde{u}_1^H\right)$, from (4.10) to (4.12), one has

$$\theta_h(0) = \left(\tilde{u}_1^H \otimes I\right) \tilde{x}(0). \tag{4.13}$$

It can be shown that

$$\begin{aligned}
&\int_0^t e^{(A+BK_1+BK_2)(t-s)}(\tilde{u}_1^H \otimes I)\dot{h}(s)ds \\
&= (\tilde{u}_1^H \otimes I)h(t)e^{(A+BK_1+BK_2)t}(\tilde{u}_1^H \otimes I)h(0) \\
&\quad + \int_0^t e^{(A+BK_1+BK_2)(t-s)}\left(\tilde{u}_1^H \otimes (ABK_1BK_2)\right) h(s)ds.
\end{aligned} \tag{4.14}$$

By (4.6), (4.13), and (4.14), the conclusion of Theorem 4.2 can be obtained.

Remark 4.6 In Theorem 4.2, $r_0(t)$ is said to be the nominal function, which describes the state formation reference function of the swarm system without external command input $v(t)$ and formation $h(t)$. $r_v(t)$ and $r_h(t)$ describe the impacts of $v(t)$ and $h(t)$, respectively. If $h(t) \equiv 0$, $r(t)$ becomes the explicit expression of the state consensus function. Moreover, from Theorem 4.2, one can see that K_2 can be used to design the motion modes of the state formation reference and the time delay has no direct effect on $r(t)$.

4.2.3 Time-Varying State Formation Feasibility Analysis and Protocol Design

In this subsection, necessary and sufficient conditions for time-varying state formation feasibility are proposed and an algorithm to design the protocol for swarm systems to achieve state formations is given.

Let $\hat{B} = [\hat{B}_1^T, \hat{B}_2^T]^T$ be a nonsingular matrix with $\hat{B}_1 \in \mathbb{R}^{m\times n}$ and $\hat{B}_2 \in \mathbb{R}^{(n-m)\times n}$ such that $\hat{B}_1 B = I$ and $\hat{B}_2 B = 0$.

Theorem 4.3 *For any given bounded initial states, a time-varying state formation $h(t)$ is feasible for swarm system (4.4) if and only if the following conditions hold simultaneously*
(i) For any $i, j \in \{1, 2, \ldots, N\}$

$$\lim_{t\to\infty} \left(\hat{B}_2 A(h_i(t) - h_j(t)) - \hat{B}_2(\dot{h}_i(t) - \dot{h}_j(t))\right) = 0; \tag{4.15}$$

(ii) The following subsystems are asymptotically stable for λ_i $(i = 2, 3, \ldots, N)$

$$\dot{\xi}(t) = (A + BK_1 + BK_2)\xi(t) - \lambda_i BK_3\xi(t - \tau). \tag{4.16}$$

Proof Necessity: If a time-varying state formation $h(t)$ is feasible for swarm system (4.4), then there exist K_i $(i = 1, 2, 3)$ and $v(t)$ such that the state formation $h(t)$ is achieved; that is,

$$\lim_{t\to\infty} (x_i(t) - h_i(t) - r(t)) = 0 \ (i = 1, 2, \ldots, N). \tag{4.17}$$

Let $\varepsilon_i(t) = (x_i(t) - h_i(t)) - (x_1(t) - h_1(t))$ $(i = 2, 3, \ldots, N)$. From (4.17), one has

$$\lim_{t\to\infty} \varepsilon_i(t) = 0. \tag{4.18}$$

Taking the derivative of $\varepsilon_i(t)$ with respect to t, one can obtain

$$\begin{aligned}\dot{\varepsilon}_i(t) = {} & (A + BK_1 + BK_2)\varepsilon_i(t) + (A + BK_1)(h_i(t) - h_1(t)) \\ & + B(v_i(t) - v_1(t)) - (\dot{h}_i(t) - \dot{h}_1(t)) \\ & + BK_3 \sum_{j=1}^{N} \left((w_{ij} - w_{1j})\varepsilon_j(t - \tau) - \deg_{in}(v_i)\varepsilon_i(t - \tau)\right).\end{aligned} \tag{4.19}$$

Let $\varepsilon(t) = [\varepsilon_2^T(t), \varepsilon_3^T(t), \ldots, \varepsilon_N^T(t)]^T$. Then

$$\begin{aligned}\dot{\varepsilon}(t) = {} & (I_{N-1} \otimes (A + BK_1 + BK_2))\,\varepsilon(t) - (I_{N-1} \otimes I)\,\dot{\bar{h}}(t) \\ & + (I_{N-1} \otimes (A + BK_1))\,\bar{h}(t) + (I_{N-1} \otimes B)\,\bar{v}(t) \\ & - \left((\bar{L} + \mathbf{1}_{N-1}\bar{w}^T) \otimes BK_3\right)\varepsilon(t - \tau),\end{aligned} \tag{4.20}$$

where

$$\bar{L} = \begin{bmatrix} \deg_{in}(v_2) & -w_{23} & \dots & -w_{2N} \\ -w_{32} & deg_{in}(v_3) & \dots & -w_{3N} \\ \vdots & \vdots & \ddots & \vdots \\ -w_{N2} & -w_{N3} & \dots & \deg_{in}(v_N) \end{bmatrix}, \ \bar{w} = \begin{bmatrix} w_{12} \\ w_{13} \\ \vdots \\ w_{1N} \end{bmatrix},$$

$$\bar{h}(t) = \begin{bmatrix} h_2(t) - h_1(t) \\ h_3(t) - h_1(t) \\ \vdots \\ h_N(t) - h_1(t) \end{bmatrix}, \ \bar{v}(t) = \begin{bmatrix} v_2(t) - v_1(t) \\ v_3(t) - v_1(t) \\ \vdots \\ v_N(t) - v_1(t) \end{bmatrix}.$$

Define

$$U_h = \begin{bmatrix} 1 & 0 \\ -\mathbf{1} & I \end{bmatrix}.$$

It can be shown that

$$U_h L U_h^{-1} = \begin{bmatrix} 0 & -w^T \\ 0 & \mathbf{1}w^T + \bar{L} \end{bmatrix}.$$

Since U_h is a nonsingular matrix and $J = \text{diag}\{0, \bar{J}\}$, there exists a nonsingular matrix $\bar{U}_h$ such that $\bar{U}_h(\mathbf{1}w^T + \bar{L})\bar{U}_h^{-1} = \bar{J}$. Let $\bar{\varepsilon}(t) = (\bar{U}_h \otimes I)\varepsilon(t)$. System (4.20) can be transformed into

$$\begin{aligned} \dot{\bar{\varepsilon}}(t) = & \left(I_{N-1} \otimes (A + BK_1 + BK_2)\right)\bar{\varepsilon}(t) - \left(\bar{J} \otimes BK_3\right)\bar{\varepsilon}(t-\tau) + \left(\bar{U}_h \otimes B\right)\bar{v}(t) \\ & + \left(\bar{U}_h \otimes (A + BK_1)\right)\bar{h}(t) - \left(\bar{U}_h \otimes I\right)\dot{\bar{h}}(t). \end{aligned} \tag{4.21}$$

To ensure that (4.18) holds for any given bounded initial states, it is required that

$$\lim_{t\to\infty}\left(\left(\bar{U}_h \otimes (A + BK_1)\right)\bar{h}(t) + \left(\bar{U}_h \otimes B\right)\bar{v}(t) - \left(\bar{U}_h \otimes I\right)\dot{\bar{h}}(t)\right) = 0, \tag{4.22}$$

and the system described by

$$\dot{\bar{\varsigma}}_h(t) = \left(I_{N-1} \otimes (A + BK_1 + BK_2)\right)\bar{\varsigma}_h(t) - \left(\bar{J} \otimes BK_3\right)\bar{\varsigma}_h(t-\tau) \tag{4.23}$$

is asymptotically stable. Since $\bar{U}_h$ is nonsingular, from (4.22), the following holds

$$\lim_{t\to\infty}\left((A + BK_1)(h_i(t) - h_j(t)) + B(v_i(t) - v_j(t)) - (\dot{h}_i(t) - \dot{h}_j(t))\right) = 0. \tag{4.24}$$

Pre-multiplying the left and right sides of (4.24) by $\hat{B}$, one can obtain that

$$\lim_{t\to\infty}\left(\hat{B}_1(A+BK_1)(h_i(t)-h_j(t))+(v_i(t)-v_j(t))-\hat{B}_1(\dot{h}_i(t)-\dot{h}_j(t))\right)=0, \quad (4.25)$$

and

$$\lim_{t\to\infty}\left(\hat{B}_2A(h_i(t)-h_j(t))-\hat{B}_2(\dot{h}_i(t)-\dot{h}_j(t))\right)=0. \quad (4.26)$$

Choosing appropriate $v_i(t)$ $(i = 1, 2, \ldots, N)$ can guarantee (4.25) holds for all $i, j \in \{1, 2, \ldots, N\}$. From (4.26), one can see that condition (i) is required. Since $\bar{J}$ consists of Jordan blocks corresponding to λ_i $(i = 2, 3, \ldots, N)$, from (4.23), it can be shown that condition (ii) is required.

Sufficiency: If condition (i) holds for state formation $h(t)$, one can obtain that

$$\lim_{t\to\infty}\left(\hat{B}_2(A+BK_1)(h_i(t)-h_j(t))-\hat{B}_2(\dot{h}_i(t)-\dot{h}_j(t))\right)=0. \quad (4.27)$$

For any $i, j \in \{1, 2, \ldots, N\}$, one can find $v_i(t)-v_j(t)$ satisfying (4.25). From (4.25) and (4.27), it holds that

$$\lim_{t\to\infty}\left(\hat{B}(A+BK_1)(h_i(t)-h_j(t))+\hat{B}B(v_i(t)-v_j(t))-\hat{B}(\dot{h}_i(t)-\dot{h}_j(t))\right)=0. \quad (4.28)$$

Pre-multiplying the left and right sides of (4.28) by $\hat{B}^{-1}$, one has

$$\lim_{t\to\infty}\left((A+BK_1)(h_i(t)-h_j(t))+B(v_i(t)-v_j(t))-(\dot{h}_i(t)-\dot{h}_j(t))\right)=0. \quad (4.29)$$

From (4.29), one can obtain

$$\lim_{t\to\infty}\left((L\otimes(A+BK_1))\,h(t)+(L\otimes B)\,v(t)-(L\otimes I)\,\dot{h}(t)\right)=0. \quad (4.30)$$

Substitute $L = UJU^{-1}$ into (4.30) and then pre-multiply the left and right sides of (4.30) by $U^{-1}\otimes I$. It follows that

$$\lim_{t\to\infty}\left((\bar{J}\tilde{U}\otimes(A+BK_1))h(t)+(\bar{J}\tilde{U}\otimes B)v(t)-(\bar{J}\tilde{U}\otimes I)\dot{h}(t)\right)=0. \quad (4.31)$$

Since the interaction topology has a spanning tree, from Lemma 2.1 and the structure of U, one knows that $\bar{J}$ is nonsingular. Pre-multiplying the left and right sides of (4.31) by $\bar{J}^{-1}\otimes I$, one has

$$\lim_{t\to\infty}\left((\tilde{U}\otimes(A+BK_1))h(t)+(\tilde{U}\otimes B)v(t)-(\tilde{U}\otimes I)\dot{h}(t)\right)=0. \quad (4.32)$$

Condition (ii) and (4.32) indicate that $\lim_{t\to\infty}\varsigma_h(t) = 0$. Then from Theorem 4.1, one sees that the time-varying state formation $h(t)$ is feasible for swarm system (4.4). This completes the proof.

Remark 4.7 Theorem 4.3 implies that the feasibility of the time-varying state formation $h(t)$ depends on the dynamics of each agent, state formation compensation signal $v(t)$, and interaction topologies. From (4.24) to (4.26), one sees that the application of $v(t)$ can expand the set of feasible state formation $h(t)$, and K_1 has no direct effect on the feasible set of $h(t)$ when $v(t)$ is applied.

Remark 4.8 If $K_2 = 0$, $h(t) \equiv 0$, $v(t) \equiv 0$, and $\tau = 0$, the time-varying state formation problems in the current section become the consensus problems discussed in [21], and Theorem 1 in [21] is just a special case of Theorem 4.3.

From Theorem 4.3, the following corollaries can be obtained directly.

Corollary 4.1 *If $v_i(t) \equiv 0$ $(i = 1, 2, \ldots, N)$, for any given bounded initial states, a time-varying state formation $h(t)$ is feasible for swarm system (4.4) if and only if condition (ii) in Theorem 4.3 holds, and for any $i, j \in \{1, 2, \ldots, N\}$,*

$$\lim_{t\to\infty}\left((A + BK_1)(h_i(t) - h_j(t)) - (\dot{h}_i(t) - \dot{h}_j(t))\right) = 0. \tag{4.33}$$

Remark 4.9 From (4.33), one can see that K_1 can be used to expand the set of feasible time-varying state formation $h(t)$ in the case where $v(t) \equiv 0$.

Corollary 4.2 *If $v_i(t) \equiv 0$ $(i = 1, 2, \ldots, N)$ and the state formation is given by a constant vector h, then for any given bounded initial states, the state formation h is feasible for swarm system (4.4) if and only if condition (ii) in Theorem 4.3 holds, and for $\forall i, j \in \{1, 2, \ldots, N\}$,*

$$(A + BK_1)(h_i - h_j) = 0. \tag{4.34}$$

Remark 4.10 In Corollary 4.2, if the state formation h is feasible for swarm system (4.4), then for all $i, j \in \{1, 2, \ldots, N\}$, $h_i - h_j$ must belong to the right null space of $A + BK_1$. If $K_1 = 0$, $K_2 = 0$, and $\tau = 0$, then Corollary 4.2 is equivalent to the Theorem 1 in Ma and Zhang [20].

Remark 4.11 Let $\kappa_i \in \mathbb{R}^n$ and $h_i = \kappa_i \otimes [1, 0]^T$ $(i = 1, 2, \ldots, N)$. If $h = [h_1^T, h_2^T, \ldots, h_N^T]^T$, $v(t) \equiv 0$, $K_1 = 0$, $K_2 = 0$, $\tau = 0$, and

$$A = \operatorname{diag}\left(\begin{bmatrix} 0 & 1 \\ a_{21}^1 & a_{22}^1 \end{bmatrix}, \ldots, \begin{bmatrix} 0 & 1 \\ a_{21}^n & a_{22}^n \end{bmatrix}\right), \quad B = I_n \otimes \begin{bmatrix} 0 \\ 1 \end{bmatrix},$$

Theorem 4.3 gives the conclusions in Lafferriere et al. [16].

Based on the above results, an algorithm to design protocol (4.3) for swarm system (4.1) to achieve time-varying state formation $h(t)$ can be summarized as follows.

Algorithm 4.1. For swarm system (4.1) with protocol (4.3), $v_i(t)$ $(i = 1, 2, \ldots, N)$ and K_i $(i = 1, 2, 3)$ can be designed in the following procedure.
Step 1: Check the feasible condition (4.15). If it is satisfied, then $v_i(t)$ $(i = 1, 2, \ldots, N)$ can be determined by Eq. (4.25), and K_1 can be any constant matrix with appropriate dimension; else time-varying state formation $h(t)$ is not feasible for swarm system (4.1) under protocol (4.3) and stop.

If it is required that $v(t) \equiv 0$, solve feasible condition (4.33) for K_1. If there exists constant gain matrix K_1 satisfying condition (4.33), then continue; else time-varying state formation $h(t)$ is not feasible for swarm system (4.1) under protocol (4.3) and stop.
Step 2: Choose K_2 to specify the motion modes of the state formation reference by assigning the eigenvalues of $A + BK_1 + BK_2$ at the desired locations in the complex plane. If (A, B) is controllable, there always exists such a K_2.
Step 3: Design K_3 to make subsystems (4.16) asymptotically stable. Many existing approaches in the literature (e.g., [22, 23]) can be used to deal with the asymptotic stability problems of subsystems (4.16) such as the approach with less calculation complexity presented in Theorem 3 in our previous work [24].

Specially, if τ=0 and (A, B) is stabilizable, then, as shown in [21], $K_3 = [\mathrm{Re}(\lambda_2)]^{-1}R_o^{-1}B^TP_o$ can make subsystems (4.16) asymptotically stable, where $P_o = P_o^T > 0$ is the solution to the algebraic Riccati equation

$$P_o(A + BK_1 + BK_2) + (A + BK_1 + BK_2)^T P_o - P_oBR_o^{-1}B^TP_o + Q_o = 0,$$

for $R_o = R_o^T > 0$ and $Q_o = D_o^TD_o \geq 0$ with $(A + BK_1 + BK_2, D_o)$ detectable.

Remark 4.12 It should be pointed out that by Eq. (4.25), $v_i(t)$ $(i = 1, 2, \ldots, N)$ cannot be uniquely determined. One can first specify a $v_i(t)$ $(i \in \{1, 2, \ldots, N\})$, then determine the other $v_j(t)$ $(j \in \{1, 2, \ldots, N\}, j \neq i)$ by Eq. (4.25).

4.2.4 Numerical Simulations

In this section, numerical examples which include a comparative example to Porfiri et al. [18] are given to illustrate the effectiveness of theoretical results obtained in this section. The interaction topologies of the swarm systems in the examples are shown in Figs. 4.2 and 4.3, respectively.

Example 4.1 Consider a swarm system with six agents. The dynamics of each agent is described by (4.1) with

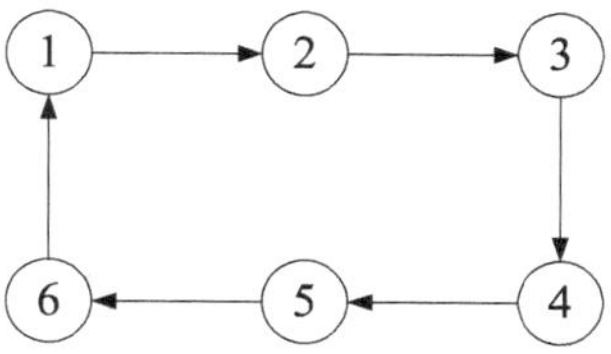

Fig. 4.2 Interaction topology G_1

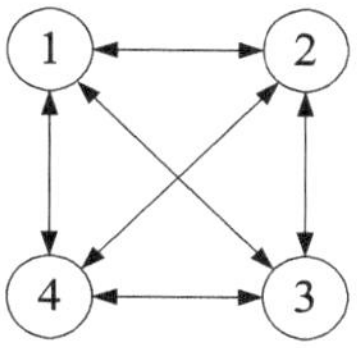

Fig. 4.3 Interaction topology G_2

$$A = \begin{bmatrix} 0 & 1 & 0 \\ 0 & 0 & 1 \\ -3 & 5 & 4 \end{bmatrix}, \; B = \begin{bmatrix} 0 \\ 0 \\ 1 \end{bmatrix}.$$

These six agents are required to preserve a periodic time-varying parallel hexagon formation and keep rotating around the predefined time-varying formation reference. The formation is defined as follows:

$$h_i(t) = \begin{bmatrix} 2\sin\left(t + \frac{(i-1)\pi}{3}\right) \\ 2\cos\left(t + \frac{(i-1)\pi}{3}\right) \\ -2\sin\left(t + \frac{(i-1)\pi}{3}\right) \end{bmatrix} \quad (i = 1, 2, \ldots, 6).$$

If the formation specified by the above $h_i(t)$ $(i = 1, 2, \ldots, 6)$ is achieved, the six agents will locate at the six diagonals of a regular hexagon, respectively, and keep rotating with an angular velocity of $1\mathrm{rad}/s$. Moreover, the edge length of the desired parallel hexagon is periodic time-varying.

Let $\tau = 0.02$ s. According to the Algorithm 4.1, K_1 can be any constant matrix with appropriate dimension, e.g., $K_1 = [-2, -4, -6]$ and for $i \in \{1, 2, \ldots, 6\}$

$$v_i(t) = 6\sin\left(t + \frac{\pi}{3}(i-1)\right) - 4\cos\left(t + \frac{\pi}{3}(i-1)\right).$$

The motion modes of the formation reference are placed at 0.02, -2.2, and -2 by $K_2 = [5.088, -5.316, -2.180]$. Using the approach in Xi et al. [24], K_3 can be obtained to make subsystems (4.16) asymptotically stable as follows:

$$K_3 = [0.4806, 0.3610, -0.0043].$$

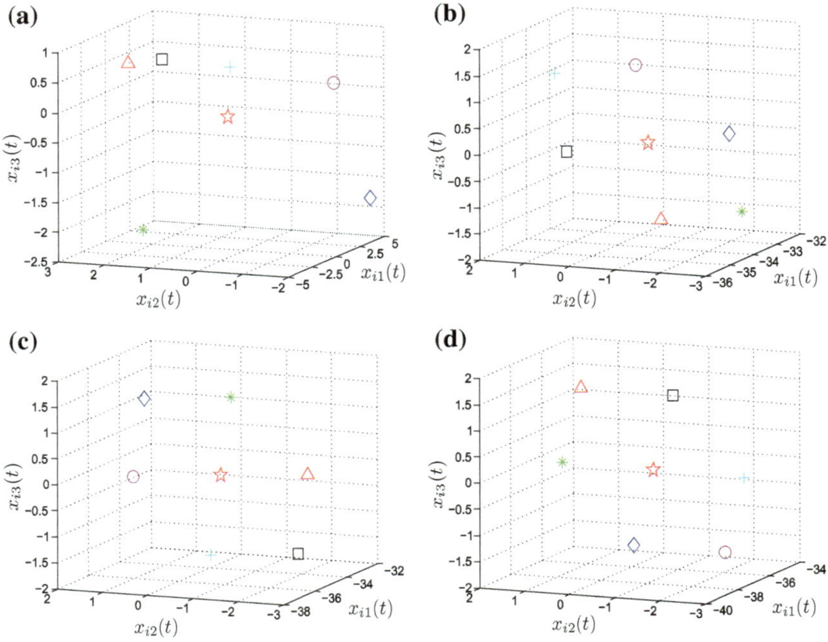

Fig. 4.4 Trajectory snapshots of six agents and $r(t)$. **a** $t = 0$ s. **b** $t = 196$ s. **c** $t = 198$ s. **d** $t = 200$ s

For simplicity of description, let the initial states of each agent be $x_{i1}(0) = 4(\Theta - 0.5)$, $x_{i2}(0) = 3(\Theta - 0.5)$, and $x_{i3}(0) = 2(\Theta - 0.5)$ where $i = 1, 2, \ldots, 6$. Figure 4.4 shows the trajectories snapshots of the six agents and the predefined formation reference at different times, where the trajectories of six agents and the predefined formation reference are denoted by the asterisk, triangle, diamond, circle, plus, square and pentagram, respectively. Figure 4.4a, b indicate that the swarm system achieves a parallel hexagon formation and the point corresponding to the predefined formation reference lies in the center of the formation. Figure 4.4b–d show that the achieved formation keeps rotating around the predefined formation reference, and both the edge length of parallel hexagon formation and the formation reference are time-varying.

Remark 4.13 For second-order swarm systems, Ren [25] and Lin and Jia [26] showed sufficient conditions to achieve rotation formation, where the agents maintain a specific formation while surrounding a common point with a constant angular velocity. By our method in Example 4.1, time-varying formation and rotation formation are achieved simultaneously.

Example 4.2 For the common virtual leader case, Porfiri et al. [18] presented a design technique for decoupling the centralized virtual leader computation from the decentralized formation control, and gave conditions for the tracking and formation stability. According to the Proposition 1 in Porfiri et al. [18], tracking and formation stability problems were reduced to the requirement that the state coefficient matrices of the reduced-order system are Hurwitz. In the following, a comparative example will be given to show that even if the state coefficient matrices of the reduced-order system are Hurwitz, the formation and trajectory tracking may not be achieved. However, using the approaches in the current section, the formation can be achieved and the motion modes of the formation reference can be specified. It should be mentioned that the interaction topology, the parameters of the vector field and the desired formation in the comparative example are all the same with those in the first simulation example in Porfiri et al. [18].

Consider a second-order swarm system with four agents moving in an affine and time-invariant vector field. The interaction topology of the swarm system is shown in Fig. 4.3. The dynamics of each agent is described by

$$\begin{cases} \dot{x}_i(t) = Ax_i(t) + Bu_i(t), \\ y_i(t) = Cx_i(t), \end{cases} \tag{4.35}$$

where $i = 1, 2, 3, 4$, $x_i(t) = [x_{i1}(t), x_{i2}(t)]^T$,

$$A = \begin{bmatrix} -1 & 2 \\ 3 & -5 \end{bmatrix},\ B = \begin{bmatrix} 1 & 0 \\ 0 & 1 \end{bmatrix},\ C = \begin{bmatrix} 1 & 0 \\ 0 & 1 \end{bmatrix}.$$

It is required that the points corresponding to states of swarm system (4.35) converge to the vertexes of a square with edge length 2 and the point corresponding to states of the virtual leader lies in the center of the square. Let $x(t) = [x_1^T(t), x_2^T(t), x_3^T(t), x_4^T(t)]^T$. The controllers for each agent and the virtual leader are the same as those adopted in the first simulation example in Porfiri et al. [18], except that

$$K_D = \begin{bmatrix} -1 & -2 \\ -3 & -5 \end{bmatrix},\ V_D = \begin{bmatrix} -2 & -1 \\ 0.5 & 0.3 \end{bmatrix}.$$

From Proposition 1 in Porfiri et al. [18], one can obtain the four state coefficient matrices of the reduced-order system as follows:

$$\Lambda_1 = \begin{bmatrix} 0 & V_D \alpha C \\ -BK_D & A + BK_D C \end{bmatrix},\ \Lambda_i = A + BK_D C(\lambda_i + 1)\ (i = 2, 3, 4),$$

where

$$\alpha = \begin{bmatrix} 0.1 & 0.02 \\ 0.02 & 0.2 \end{bmatrix},$$

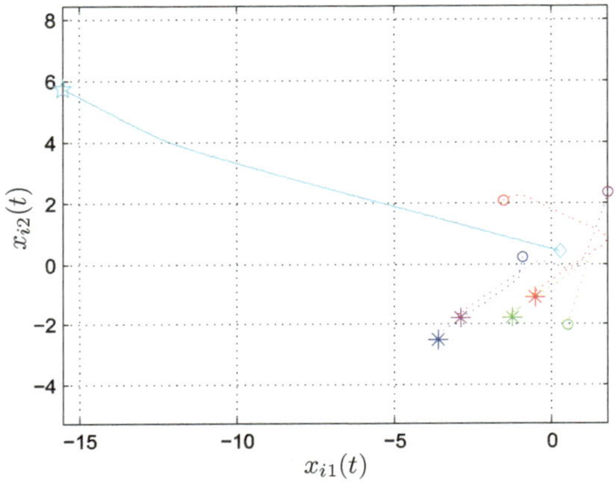

Fig. 4.5 State trajectories of four agents and the virtual leader using the approach of Porfiri et al. [18]

and λ_i $(i = 2, 3, 4)$ are the eigenvalues of L with $\lambda_2 = \lambda_3 = \lambda_4 = 4$. It can be verified that Λ_i $(i = 1, 2, 3, 4)$ are Hurwitz. Therefore, according to the results of Porfiri et al. [18], the tracking and formation stabilities can be guaranteed. Let $q(0) = [\Theta - 0.5, \Theta - 0.5]^T$, $x_i(0) = [4(\Theta - 0.5), 5(\Theta - 0.5)]^T$ $(i = 1, 2, 3, 4)$. Figure 4.5 shows the state trajectories of the four agents and the virtual leader within $20s$, where the trajectories of agents and virtual leader are marked by dotted lines and solid line, respectively. Moreover, the initial states of agents and virtual leader are denoted by circles and diamond, respectively, and the final states of agents and virtual leader are denoted by asterisks and pentagram, respectively. From Fig. 4.5, one can see that neither the predefined formation nor the tracking is achieved.

In the following, we will show that using the results in the current section, not only the desired square formation can be achieved but also the motion modes of the formation reference can be specified.

Let $v_i(t) \equiv 0$, $\tau = 0$. From the Algorithm 4.1, one has

$$K_1 = \begin{bmatrix} 1 & -2 \\ -3 & 5 \end{bmatrix}.$$

The formation reference can be designed to move in a circle by placing the motion modes of the formation reference at j and $-j$ with $j^2 = -1$ by

$$K_2 = \begin{bmatrix} 0 & -1 \\ 1 & 0 \end{bmatrix}.$$

Using the approach in the Algorithm 4.1, one can obtain a K_3 to make subsystems (4.16) asymptotically stable as follows:

$$K_3 = \begin{bmatrix} 0.14 & 0.0429 \\ 0.0429 & 0.198 \end{bmatrix}.$$

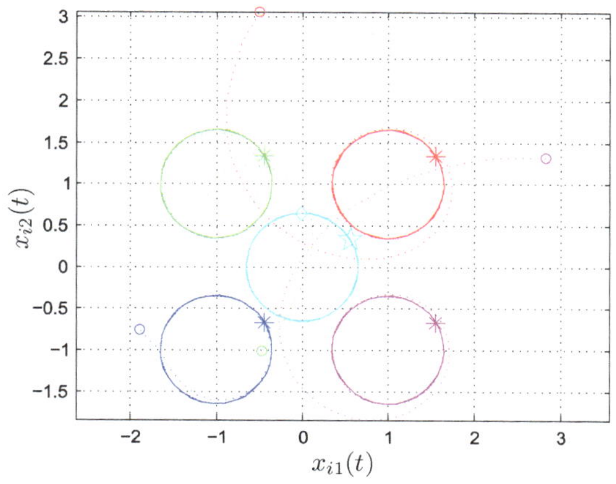

Fig. 4.6 State trajectories of four agents and $r(t)$ using the approach in this section

Figure 4.6 shows the state trajectories of the four agents and the predefined formation reference within 20 s. From Fig. 4.6, one can see that the desired square formation is achieved and the formation reference moves on a circle.

4.3 Time-Varying Output Formation Control

This section studies time-varying output formation control problems for high-order LTI swarm systems. First, a time-varying output formation control protocol is presented based on the static output feedback approach. Then using output transformation, the time-varying output formation problem is converted into the output consensus problem. Based on the output space decomposition approach and partial stability theory, necessary and sufficient conditions for swarm systems to achieve time-varying output formations and an explicit expression of the output formation reference function are proposed. Approaches to partially assign the motion modes of the output formation reference are given. Furthermore, necessary and sufficient conditions for output formation feasibilities are presented. Finally, approaches to expand the set of output formation feasibility and approaches to design the time-varying output formation protocol are given.

4.3.1 Problem Description

Consider the following swarm system consisting of N agents

$$\begin{cases} \dot{x}_i(t) = Ax_i(t) + Bu_i(t), \\ y_i(t) = Cx_i(t), \end{cases} \tag{4.36}$$

where $i = 1, 2, \ldots, N$, $x_i(t) \in \mathbb{R}^n$, $u_i(t) \in \mathbb{R}^m$, $y_i(t) \in \mathbb{R}^q$ denote the state, control input, and output of agent i, respectively. It is assumed that $\text{rank}(C) = q$ and the interaction topology has a spanning tree. A time-varying output formation is specified by a vector $h(t) = [h_1^T(t), h_2^T(t), \ldots, h_N^T(t)]^T \in \mathbb{R}^{Nq}$.

Definition 4.5 Swarm system (4.36) is said to *achieve time-varying output formation $h(t)$* if for any given bounded initial states, there exists a vector-valued function $r(t) \in \mathbb{R}^q$ such that

$$\lim_{t\to\infty} (y_i(t) - h_i(t) - r(t)) = 0 \ (i = 1, 2, \ldots, N), \tag{4.37}$$

where $r(t)$ is called an *output formation reference function*.

Definition 4.6 If there exist control inputs $u_i(t)$ $(i = 1, 2, \ldots, N)$ such that for any given bounded initial states, swarm system (4.36) achieves time-varying output formation $h(t)$, then $h(t)$ is said to be *feasible* for swarm system (4.36).

Definition 4.7 Swarm system (4.36) is said to *achieve output consensus* if for any given bounded initial states, there exists a vector-valued function $c(t) \in \mathbb{R}^q$ such that

$$\lim_{t\to\infty} (y_i(t) - c(t)) = 0 \ (i = 1, 2, \ldots, N),$$

where $c(t)$ is called an *output consensus function*.

Remark 4.14 From Definitions 4.5 and 4.7, one sees that if $h(t) \equiv 0$, the output formation reference function is equivalent to the output consensus function. If $C = I$ or $h(t) \equiv 0$, output formation problems become state formation problems or output consensus problems. Therefore, state formation problems and output consensus problems can be regarded as special cases of output formation problems.

Consider a given time-varying output formation $h(t)$ with $h_i(t)$ $(i = 1, 2, \ldots, N)$ piecewise continuously differentiable and consider the following output formation protocol.

$$\begin{aligned} u_i(t) = {} & K_1 y_i(t) + K_2 (y_i(t) - h_i(t)) \\ & + K_3 \sum_{j\in N_i} w_{ij} \left((y_i(t) - h_i(t)) - (y_j(t) - h_j(t))\right) + v_i(t), \end{aligned} \tag{4.38}$$

where $i = 1, 2, \ldots, N$, K_1, K_2, and K_3 are constant gain matrices with appropriate dimensions, and $v_i(t) \in \mathbb{R}^m$ is the output formation compensation signal to be determined later.

Remark 4.15 In protocol (4.38), K_1 and $v_i(t)$ $(i = 1, 2, \ldots, N)$ can be used to expand the feasible output formation set. K_2 can be used to partially assign the motion modes of the output formation reference. The term with K_3 is used to drive the outputs of all agents to achieve the desired time-varying output formation. It should be pointed out that K_1, K_2, and $v_i(t)$ $(i = 1, 2, \ldots, N)$ are not necessary for swarm systems to achieve some output formations. If $K_1 = 0$, $K_2 = 0$, and $v_i(t) \equiv 0$, protocol (4.38) becomes the one that only uses relative information of neighbors.

Definition 4.8 A time-varying output formation $h(t)$ is *feasible* for swarm system (4.36) under protocol (4.38), if there exist K_i $(i = 1, 2, 3)$ and $v_i(t)$ $(i = 1, 2, \ldots, N)$ such that it can be achieved for any given bounded initial states.

Let $y(t) = [y_1^T(t), y_2^T(t), \ldots, y_N^T(t)]^T$. Under protocol (4.38), swarm system (4.36) can be written in a compact form as follows:

$$\begin{cases} \dot{x}(t) = (I_N \otimes (A + BK_1C + BK_2C) + L \otimes BK_3C)\, x(t) \\ \qquad\qquad - (I_N \otimes BK_2 + L \otimes BK_3)\, h(t) + (I_N \otimes B)\, v(t), \\ y(t) = (I_N \otimes C)\, x(t). \end{cases} \tag{4.39}$$

In the current section, the following three problems for swarm system (4.39) are mainly addressed: (i) under what conditions the time-varying output formation $h(t)$ can be achieved; (ii) under what conditions a time-varying output formation $h(t)$ is feasible; and (iii) how to design protocol (4.38) to achieve time-varying output formation $h(t)$.

4.3.2 Time-Varying Output Formation Analysis

In this subsection, first, a consensus-based approach is introduced to deal with the time-varying output formation problems. Then, necessary and sufficient conditions for swarm system (4.39) to achieve time-varying output formation $h(t)$ are presented, and an explicit expression of the output formation reference function is given.

Since C is of full row rank, there exists $\bar{C} \in \mathbb{R}^{(n-q)\times n}$ such that $T = [C^T, \bar{C}^T]^T$ is nonsingular. Let $\bar{y}_i(t) = \bar{C}x_i(t)$ $(i = 1, 2, \ldots, N)$, $\bar{y}(t) = [\bar{y}_1^T(t), \bar{y}_2^T(t), \ldots, \bar{y}_N^T(t)]^T$ and

$$TAT^{-1} = \begin{bmatrix} \bar{A}_{11} & \bar{A}_{12} \\ \bar{A}_{21} & \bar{A}_{22} \end{bmatrix}, \quad TB = \begin{bmatrix} \bar{B}_1 \\ \bar{B}_2 \end{bmatrix}.$$

Applying the nonsingular transformation $I_N \otimes T$ to swarm system (4.39) leads to

$$\begin{cases} \dot{y}(t) = \left(I_N \otimes \left(\bar{A}_{11} + \bar{B}_1K_1 + \bar{B}_1K_2\right) + L \otimes \bar{B}_1K_3\right) y(t) + \left(I_N \otimes \bar{A}_{12}\right) \bar{y}(t) \\ \qquad - \left(I_N \otimes \bar{B}_1K_2 + L \otimes \bar{B}_1K_3\right) h(t) + \left(I_N \otimes \bar{B}_1\right) v(t), \\ \dot{\bar{y}}(t) = \left(I_N \otimes \left(\bar{A}_{21} + \bar{B}_2K_1 + \bar{B}_2K_2\right) + L \otimes \bar{B}_2K_3\right) y(t) + \left(I_N \otimes \bar{A}_{22}\right) \bar{y}(t) \\ \qquad - \left(I_N \otimes \bar{B}_2K_2 + L \otimes \bar{B}_2K_3\right) h(t) + \left(I_N \otimes \bar{B}_2\right) v(t). \end{cases} \tag{4.40}$$

The two subsystems with states $y(t)$ and $\bar{y}(t)$ describe the output component and non-output component of swarm system (4.39), respectively. Let $\tilde{y}_i(t) = y_i(t) - h_i(t)$ $(i = 1, 2, \ldots, N)$, and $\tilde{y}(t) = [\tilde{y}_1^T(t), \tilde{y}_2^T(t), \ldots, \tilde{y}_N^T(t)]^T$, then swarm system (4.40) can be converted into

$$\begin{cases} \dot{\tilde{y}}(t) = \left(I_N \otimes \left(\bar{A}_{11} + \bar{B}_1 K_1 + \bar{B}_1 K_2\right) + L \otimes \bar{B}_1 K_3\right) \tilde{y}(t) + \left(I_N \otimes \bar{A}_{12}\right) \bar{y}(t) \\ \qquad + \left(I_N \otimes \left(\bar{A}_{11} + \bar{B}_1 K_1\right)\right) h(t) - (I_N \otimes I)\, \dot{h}(t) + \left(I_N \otimes \bar{B}_1\right) v(t), \\ \dot{\bar{y}}(t) = \left(I_N \otimes \left(\bar{A}_{21} + \bar{B}_2 K_1 + \bar{B}_2 K_2\right) + L \otimes \bar{B}_2 K_3\right) \tilde{y}(t) + \left(I_N \otimes \bar{A}_{22}\right) \bar{y}(t) \\ \qquad + \left(I_N \otimes \left(\bar{A}_{21} + \bar{B}_2 K_1\right)\right) h(t) + \left(I_N \otimes \bar{B}_2\right) v(t). \end{cases} \tag{4.41}$$

Choose $\tilde{y}(t)$ as the output of swarm system (4.41). Then the following lemma holds directly.

Lemma 4.2 *For any given bounded initial states, swarm system (4.39) achieves time-varying output formation $h(t)$ if and only if swarm system (4.41) achieves output consensus.*

Choose the v in Sect. 2.2 as $v = q$. Let $\theta_y(t) = (\tilde{u}_1^H \otimes I)\tilde{y}(t)$, $\varsigma_y(t) = (\tilde{U} \otimes I)\tilde{y}(t)$, $\bar{\theta}_y(t) = (\tilde{u}_1^H \otimes I)\bar{y}(t)$, and $\bar{\varsigma}_y(t) = (\tilde{U} \otimes I)\bar{y}(t)$, then swarm system (4.41) can be transformed into the following two subsystems:

$$\begin{cases} \dot{\theta}_y(t) = \left(\bar{A}_{11} + \bar{B}_1 K_1 + \bar{B}_1 K_2\right) \theta_y(t) + \bar{A}_{12} \bar{\theta}_y(t) + \left(\tilde{u}_1^H \otimes \left(\bar{A}_{11} + \bar{B}_1 K_1\right)\right) h(t) \\ \qquad - \left(\tilde{u}_1^H \otimes I\right) \dot{h}(t) + \left(\tilde{u}_1^H \otimes \bar{B}_1\right) v(t), \\ \dot{\bar{\theta}}_y(t) = \left(\bar{A}_{21} + \bar{B}_2 K_1 + \bar{B}_2 K_2\right) \theta_y(t) + \bar{A}_{22} \bar{\theta}_y(t) + \left(\tilde{u}_1^H \otimes \left(\bar{A}_{21} + \bar{B}_2 K_1\right)\right) h(t) \\ \qquad + \left(\tilde{u}_1^H \otimes \bar{B}_2\right) v(t), \end{cases} \tag{4.42}$$

$$\begin{cases} \dot{\varsigma}_y(t) = \left(I_{N-1} \otimes \left(\bar{A}_{11} + \bar{B}_1 K_1 + \bar{B}_1 K_2\right) + \bar{J} \otimes \bar{B}_1 K_3\right) \varsigma_y(t) + \left(I_{N-1} \otimes \bar{A}_{12}\right) \bar{\varsigma}_y(t) \\ \qquad + \left(\tilde{U} \otimes \left(\bar{A}_{11} + \bar{B}_1 K_1\right)\right) h(t) - \left(\tilde{U} \otimes I\right) \dot{h}(t) + \left(\tilde{U} \otimes \bar{B}_1\right) v(t), \\ \dot{\bar{\varsigma}}_y(t) = \left(I_{N-1} \otimes \left(\bar{A}_{21} + \bar{B}_2 K_1 + \bar{B}_2 K_2\right) + \bar{J} \otimes \bar{B}_2 K_3\right) \varsigma_y(t) + \left(I_{N-1} \otimes \bar{A}_{22}\right) \bar{\varsigma}_y(t) \\ \qquad + \left(\tilde{U} \otimes \left(\bar{A}_{21} + \bar{B}_2 K_1\right)\right) h(t) + \left(\tilde{U} \otimes \bar{B}_2\right) v(t). \end{cases} \tag{4.43}$$

A necessary and sufficient condition for swarm system (4.39) to achieve time-varying output formation $h(t)$ is presented as follows:

Theorem 4.4 *For any given bounded initial states, swarm system (4.39) achieves time-varying output formation $h(t)$ if and only if*

$$\lim_{t \to \infty} \varsigma_y(t) = 0.$$

Proof Let $\tilde{y}_C(t) = (U \otimes I)[\theta_y^H(t), 0]^H$. Note that $[\theta_y^H(t), 0]^H = e_1 \otimes \theta_y(t)$, so there exist $\alpha_r(t)$ $(r = 1, 2, \ldots, q)$ such that

$$\tilde{y}_C(t) = \mathbf{1} \otimes \theta_y(t) = \sum_{r=1}^{q} \alpha_r(t) p_r \in \mathbb{C}(U). \tag{4.44}$$

Similarly, let $\tilde{y}_{\bar{C}}(t) = (U \otimes I)[0, \varsigma_y^H(t)]^H$, then there exist $\alpha_j(t)$ $(j = q+1, q+2, \ldots, qN)$ such that

$$\tilde{y}_{\bar{C}}(t) = \sum_{i=2}^{N} \sum_{r=1}^{q} \alpha_{(i-1)q+r}(t)(e_i \otimes c_r) = \sum_{j=q+1}^{Nq} \alpha_j(t) p_j \in \bar{\mathbb{C}}(U). \tag{4.45}$$

Since $[\theta_y^H(t), \varsigma_y^H(t)]^H = (U^{-1} \otimes I)\tilde{y}(t)$, one knows

$$\tilde{y}(t) = \tilde{y}_C(t) + \tilde{y}_{\bar{C}}(t). \tag{4.46}$$

From Lemmas 4.2, 2.3 and (4.44)–(4.46), one sees that for any given bounded initial states, swarm system (4.39) achieves time-varying output formation $h(t)$ if and only if $\lim_{t\to\infty} \tilde{y}_{\bar{C}}(t) = 0$ or $\lim_{t\to\infty} \varsigma_y(t) = 0$ as $U \otimes I$ is nonsingular. This completes the proof.

In the sequel, an explicit expression of the output formation reference function is presented.

Theorem 4.5 *If swarm system (4.39) achieves time-varying output formation $h(t)$ with the output formation reference function $r(t)$, then*

$$\lim_{t\to\infty} (r(t) - r_0(t) - r_v(t) - r_h(t)) = 0,$$

where

$$r_0(t) = Ce^{(A+BK_1C+BK_2C)t} \left(\tilde{u}_1^H \otimes I\right) x(0),$$

$$r_v(t) = C \int_0^t e^{(A+BK_1C+BK_2C)(t-\tau)} B \left(\tilde{u}_1^H \otimes I\right) v(\tau) d\tau,$$

$$r_h(t) = -\left(\tilde{u}_1^H \otimes I\right) h(t) - C \int_0^t e^{(A+BK_1C+BK_2C)(t-\tau)} BK_2 \left(\tilde{u}_1^H \otimes I\right) h(\tau) d\tau.$$

Proof From Theorem 4.4, one sees that if swarm system (4.39) achieves time-varying output formation $h(t)$, then

$$\lim_{t\to\infty} (\tilde{y}(t) - \tilde{y}_C(t)) = \lim_{t\to\infty} \left(\tilde{y}(t) - \mathbf{1} \otimes \theta_y(t)\right) = 0, \tag{4.47}$$

that is, the output formation reference function is determined by subsystem (4.42). Let $\tilde{\theta}_y(t) = [\theta_y^H(t), \bar{\theta}_y^H(t)]^H$, subsystem (4.42) can be rewritten as

$$
\begin{aligned}
\dot{\tilde{\theta}}_y(t) &= T\left(A + BK_1C + BK_2C\right)T^{-1}\tilde{\theta}_y(t) \\
&\quad + T\left(\tilde{u}_1^H \otimes AT^{-1}\begin{bmatrix} I \\ 0 \end{bmatrix} + \tilde{u}_1^H \otimes BK_1\right)h(t) \\
&\quad + \left(\tilde{u}_1^H \otimes TB\right)v(t) - \left(\tilde{u}_1^H \otimes \begin{bmatrix} I \\ 0 \end{bmatrix}\right)\dot{h}(t).
\end{aligned}
\tag{4.48}
$$

It holds that

$$
\begin{aligned}
&\int_0^t e^{(A+BK_1C+BK_2C)(t-\tau)}\left(\tilde{u}_1^H \otimes \begin{bmatrix} I \\ 0 \end{bmatrix}\right)\dot{h}(\tau)d\tau \\
&= \left(\tilde{u}_1^H \otimes \begin{bmatrix} I \\ 0 \end{bmatrix}\right)h(t) - e^{(A+BK_1C+BK_2C)t}\left(\tilde{u}_1^H \otimes \begin{bmatrix} I \\ 0 \end{bmatrix}\right)h(0) \\
&\quad + \int_0^t e^{(A+BK_1C+BK_2C)(t-\tau)}\left(\tilde{u}_1^H \otimes (A + BK_1C + BK_2C)\begin{bmatrix} I \\ 0 \end{bmatrix}\right)h(\tau)d\tau.
\end{aligned}
\tag{4.49}
$$

It can be shown that $\theta_y(0) = (\tilde{u}_1^H \otimes I)(y(0) - h(0))$ and $\bar{\theta}_y(0) = (\tilde{u}_1^H \otimes I)\bar{y}(0)$. Since $y(0) = (I \otimes C)x(0)$ and $\bar{y}(0) = (I \otimes \bar{C})x(0)$, one has

$$
\theta_y(0) = (\tilde{u}_1^H \otimes C)x(0) - (\tilde{u}_1^H \otimes I)h(0), \tag{4.50}
$$

$$
\bar{\theta}_y(0) = (\tilde{u}_1^H \otimes \bar{C})x(0). \tag{4.51}
$$

From (4.47)–(4.51), the conclusion of Theorem 4.5 can be obtained.

Remark 4.16 In Theorem 4.5, $r_0(t)$ is called the output consensus function. $r_v(t)$ and $r_h(t)$ describe the impacts of $v(t)$ and $h(t)$, respectively. If $h(t) \equiv 0$ (or $h(t) \equiv 0$ and $C = I$), $r(t)$ becomes the explicit expression of the output consensus function (or state consensus function). Moreover, from Theorem 4.5, one sees that both K_1 and K_2 can be used to assign partially the motion modes of the output formation reference. However, since K_1 is designed to expand the feasible output formation set, only K_2 has this function.

4.3.3 Time-Varying Output Formation Feasibility Analysis and Protocol Design

In this subsection, necessary and sufficient conditions for output formation feasibility are proposed and an algorithm to design the protocol for swarm systems to achieve time-varying output formation is given.

In (4.42) and (4.43), the subsystems with states $\theta_y(t)$ and $\varsigma_y(t)$ can be used to determine the output consensus and complement output consensus parts of swarm system (4.41), respectively. It can be seen that only observable components of $(\bar{A}_{22}, \bar{A}_{12})$ influence the subsystem with state $\varsigma_y(t)$. Therefore the observability decomposition of $(\bar{A}_{22}, \bar{A}_{12})$ is presented first as follows. Let $\tilde{T}$ be a nonsingular matrix such that

$$\left(\tilde{T}^{-1}\bar{A}_{22}\tilde{T}, \bar{A}_{12}\tilde{T}\right) = \left(\begin{bmatrix} \tilde{D}_1 & 0 \\ \tilde{D}_2 & \tilde{D}_3 \end{bmatrix}, \begin{bmatrix} \tilde{E}_1 & 0 \end{bmatrix}\right)$$

where $\tilde{D}_1 \in \mathbb{R}^g$ and $(\tilde{D}_1, \tilde{E}_1)$ is completely observable. Let $\tilde{\varsigma}_i(t) = \tilde{T}^{-1}\bar{\varsigma}_{yi}(t) = [\tilde{\varsigma}_{io}^T(t), \tilde{\varsigma}_{i\bar{o}}^T(t)]^T$ $(i = 1, 2, \ldots, N)$, $\tilde{\varsigma}_{yo}(t) = [\tilde{\varsigma}_{1o}^T(t), \tilde{\varsigma}_{2o}^T(t), \ldots, \tilde{\varsigma}_{No}^T(t)]^T$, $\tilde{\varsigma}_{y\bar{o}}(t) = [\tilde{\varsigma}_{1\bar{o}}^T(t), \tilde{\varsigma}_{2\bar{o}}^T(t), \ldots, \tilde{\varsigma}_{N\bar{o}}^T(t)]^T$, $\tilde{T}^{-1}\bar{A}_{21} = [\tilde{F}_1^T, \tilde{F}_2^T]^T$ and $\tilde{T}^{-1}\bar{B}_2 = [\tilde{B}_1^T, \tilde{B}_2^T]^T$. Then system (4.43) can be transformed into

$$\begin{cases} \dot{\varsigma}_y(t) = \left(I_{N-1} \otimes (\bar{A}_{11} + \bar{B}_1K_1 + \bar{B}_1K_2) + \bar{J} \otimes \bar{B}_1K_3\right)\varsigma_y(t) + \left(I_{N-1} \otimes \tilde{E}_1\right)\tilde{\varsigma}_{yo}(t) \\ \quad + \left(\tilde{U} \otimes (\bar{A}_{11} + \bar{B}_1K_1)\right)h(t) - \left(\tilde{U} \otimes I\right)\dot{h}(t) + \left(\tilde{U} \otimes \bar{B}_1\right)v(t), \\ \dot{\tilde{\varsigma}}_{yo}(t) = \left(I_{N-1} \otimes \left(\tilde{F}_1 + \tilde{B}_1K_1 + \tilde{B}_1K_2\right) + \bar{J} \otimes \tilde{B}_1K_3\right)\varsigma_y(t) + \left(I_{N-1} \otimes \tilde{D}_1\right)\tilde{\varsigma}_{yo}(t) \\ \quad + \left(\tilde{U} \otimes \left(\tilde{F}_1 + \tilde{B}_1K_1\right)\right)h(t) + \left(\tilde{U} \otimes \tilde{B}_1\right)v(t), \\ \dot{\tilde{\varsigma}}_{y\bar{o}}(t) = \left(I_{N-1} \otimes \left(\tilde{F}_2 + \tilde{B}_2K_1 + \tilde{B}_2K_2\right) + \bar{J} \otimes \tilde{B}_2K_3\right)\varsigma_y(t) + \left(I_{N-1} \otimes \tilde{D}_2\right)\tilde{\varsigma}_{yo}(t) \\ \quad + \left(I_{N-1} \otimes \tilde{D}_3\right)\tilde{\varsigma}_{y\bar{o}}(t) + \left(\tilde{U} \otimes \left(\tilde{F}_2 + \tilde{B}_2K_1\right)\right)h(t) + \left(\tilde{U} \otimes \tilde{B}_2\right)v(t). \end{cases} \tag{4.52}$$

In (4.52), the subsystems with states $\tilde{\varsigma}_{yo}(t)$ and $\tilde{\varsigma}_{y\bar{o}}(t)$ describe the observable component and nonobservable component of $\bar{\varsigma}_y(t)$, and only $\tilde{\varsigma}_{yo}(t)$ has effect on $\varsigma_y(t)$. For $\bar{B}_1$ and $\tilde{B}_1$, there exist nonsingular matrices $\bar{T} = \begin{bmatrix} \bar{T}_{11} & \bar{T}_{12} \\ \bar{T}_{21} & \bar{T}_{22} \end{bmatrix}$ and $\hat{T}$, where $\hat{T}^{-1} = \begin{bmatrix} \hat{F}_1 \\ \hat{F}_2 \end{bmatrix}$ such that

$$\bar{T}\begin{bmatrix} \bar{B}_1 \\ \tilde{B}_1 \end{bmatrix}\hat{T} = \begin{bmatrix} I & 0 \\ 0 & 0 \end{bmatrix}. \tag{4.53}$$

Theorem 4.6 *For any given bounded initial states, a time-varying output formation* $h(t)$ *is feasible for swarm system (4.39) if and only if the following conditions hold simultaneously*
(i) For any $i \in \{1, 2, \ldots, N\}$ *and* $j \in N_i$,

$$\lim_{t \to \infty}\left(\left(\bar{T}_{21}\bar{A}_{11} + \bar{T}_{22}\tilde{F}_1\right)\left(h_i(t) - h_j(t)\right) - \bar{T}_{21}\left(\dot{h}_i(t) - \dot{h}_j(t)\right)\right) = 0;$$

(ii) The following $N - 1$ *matrices are Hurwitz*

$$\bar{\Upsilon}_i = \begin{bmatrix} \bar{A}_{11} + \bar{B}_1K_1 + \bar{B}_1K_2 + \lambda_i\bar{B}_1K_3 & \tilde{E}_1 \\ \tilde{F}_1 + \tilde{B}_1K_1 + \tilde{B}_1K_2 + \lambda_i\tilde{B}_1K_3 & \tilde{D}_1 \end{bmatrix} \ (i = 2, 3, \ldots, N). \tag{4.54}$$

Proof Necessity: If a time-varying output formation $h(t)$ is feasible for swarm system (4.39), then there exist K_i $(i = 1, 2, 3)$ and $v(t)$ such that the output formation $h(t)$ is achieved. From Theorem (4.4), one has

$$\lim_{t\to\infty} \varsigma_y(t) = 0. \tag{4.55}$$

By (4.52) and Lemma 2.12, if (4.55) holds for any given bounded initial states, then it is required that

$$\lim_{t\to\infty}\left(\begin{bmatrix} \tilde{U}\otimes(\bar{A}_{11}+\bar{B}_1K_1) \\ \tilde{U}\otimes\left(\tilde{F}_1+\tilde{B}_1K_1\right)\end{bmatrix} h(t) - \begin{bmatrix}\tilde{U}\otimes I \\ 0\end{bmatrix}\dot{h}(t) + \begin{bmatrix}\tilde{U}\otimes\bar{B}_1 \\ \tilde{U}\otimes\tilde{B}_1\end{bmatrix} v(t)\right) = 0, \tag{4.56}$$

and the system described by

$$\dot{\xi}_y(t) = \begin{bmatrix} I_{N-1}\otimes(\bar{A}_{11}+\bar{B}_1K_1+\bar{B}_1K_2)+\bar{J}\otimes\bar{B}_1K_3 & I_{N-1}\otimes\tilde{E}_1 \\ I_{N-1}\otimes\left(\tilde{F}_1+\tilde{B}_1K_1+\tilde{B}_1K_2\right)+\bar{J}\otimes\tilde{B}_1K_3 & I_{N-1}\otimes\tilde{D}_1 \end{bmatrix}\xi_y(t) \tag{4.57}$$

is asymptotically stable.

Let $\tilde{U} = [\hat{u}_1, \hat{u}_2, \ldots, \hat{u}_N]$ with $\hat{u}_i \in \mathbb{R}^{(N-1)\times 1}$ $(i = 1, 2, \ldots, N)$. Since $\tilde{U}\mathbf{1}_N = 0$ and $\tilde{U}$ is of full row rank, one can find $k \in \{1, 2, \ldots, N\}$ such that

$$\hat{u}_k = -\hat{U}\mathbf{1}_{N-1}, \tag{4.58}$$

where $\hat{U} = [\hat{u}_1, \hat{u}_2, \ldots, \hat{u}_{k-1}, \hat{u}_{k+1}, \ldots, \hat{u}_N]$. Let $\bar{h}_y(t) = [h_1^T(t), h_2^T(t), \ldots, h_{k-1}^T(t), h_{k+1}^T(t), \ldots h_N^T(t)]^T$ and $\bar{v}_y(t) = [v_1^T(t), v_2^T(t), \ldots, v_{k-1}^T(t), v_{k+1}^T(t), \ldots v_N^T(t)]^T$. From (4.56) and (4.58), one has

$$\lim_{t\to\infty}\left(\hat{U}\otimes I\right)\left(\Phi_{\bar{h}_y} - \Phi_{h_k}\right) = 0, \tag{4.59}$$

where

$$\begin{aligned}\Phi_{\bar{h}_y} &= \left(I_{N-1}\otimes\begin{bmatrix}\bar{A}_{11}+\bar{B}_1K_1 \\ \tilde{F}_1+\tilde{B}_1K_1\end{bmatrix}\right)\bar{h}_y(t) - \left(I_{N-1}\otimes\begin{bmatrix} I \\ 0\end{bmatrix}\right)\dot{\bar{h}}_y(t) \\ &\quad + \left(I_{N-1}\otimes\begin{bmatrix}\bar{B}_1 \\ \tilde{B}_1\end{bmatrix}\right)\bar{v}_y(t),\end{aligned}$$

$$\begin{aligned}\Phi_{h_k} &= \left(\mathbf{1}_{N-1}\otimes\begin{bmatrix}\bar{A}_{11}+\bar{B}_1K_1 \\ \tilde{F}_1+\tilde{B}_1K_1\end{bmatrix}\right)h_k(t) - \left(\mathbf{1}_{N-1}\otimes\begin{bmatrix} I \\ 0\end{bmatrix}\right)\dot{h}_k(t) \\ &\quad + \left(\mathbf{1}_{N-1}\otimes\begin{bmatrix}\bar{B}_1 \\ \tilde{B}_1\end{bmatrix}\right)v_k(t).\end{aligned}$$

Pre-multiplying both sides of (4.59) by $\hat{U}^{-1} \otimes I$, one has that for any $i \in \{1, 2, \ldots, k-1, k+1, \ldots, N\}$,

$$\lim_{t\to\infty}\left(\begin{bmatrix}\bar{A}_{11}+\bar{B}_1K_1\\ \tilde{F}_1+\tilde{B}_1K_1\end{bmatrix}(h_i(t)-h_k(t)) - \begin{bmatrix}I\\0\end{bmatrix}\left(\dot{h}_i(t)-\dot{h}_k(t)\right)+\begin{bmatrix}\bar{B}_1\\ \tilde{B}_1\end{bmatrix}(v_i(t)-v_k(t))\right)=0. \tag{4.60}$$

which means that for any $i \in \{1, 2, \ldots, N\}$ and $j \in N_i$,

$$\lim_{t\to\infty}\left(\begin{bmatrix}\bar{A}_{11}+\bar{B}_1K_1\\ \tilde{F}_1+\tilde{B}_1K_1\end{bmatrix}\left(h_i(t)-h_j(t)\right) - \begin{bmatrix}I\\0\end{bmatrix}\left(\dot{h}_i(t)-\dot{h}_j(t)\right)+\begin{bmatrix}\bar{B}_1\\ \tilde{B}_1\end{bmatrix}\left(v_i(t)-v_j(t)\right)\right)=0. \tag{4.61}$$

From (4.53), one gets

$$\bar{T}\begin{bmatrix}\bar{B}_1\\ \tilde{B}_1\end{bmatrix}=\begin{bmatrix}\hat{F}_1\\0\end{bmatrix}. \tag{4.62}$$

Pre-multiplying both sides of (4.61) by $\bar{T}$ yields

$$\lim_{t\to\infty}\left(\left(\bar{T}\begin{bmatrix}\bar{A}_{11}\\ \tilde{F}_1\end{bmatrix}+\begin{bmatrix}\hat{F}_1K_1\\0\end{bmatrix}\right)\left(h_i(t)-h_j(t)\right) - \bar{T}\begin{bmatrix}I\\0\end{bmatrix}\left(\dot{h}_i(t)-\dot{h}_j(t)\right)+\begin{bmatrix}\hat{F}_1\\0\end{bmatrix}\left(v_i(t)-v_j(t)\right)\right)=0. \tag{4.63}$$

Substituting $\bar{T}=\begin{bmatrix}\bar{T}_{11} & \bar{T}_{12}\\ \bar{T}_{21} & \bar{T}_{22}\end{bmatrix}$ into (4.63), one can obtain that

$$\lim_{t\to\infty}\left(\left(\bar{T}_{11}\bar{A}_{11}+\bar{T}_{12}\tilde{F}_1+\hat{F}_1K_1\right)\left(h_i(t)-h_j(t)\right) - \bar{T}_{11}\left(\dot{h}_i(t)-\dot{h}_j(t)\right)+\hat{F}_1\left(v_i(t)-v_j(t)\right)\right)=0, \tag{4.64}$$

and

$$\lim_{t\to\infty}\left(\left(\bar{T}_{21}\bar{A}_{11}+\bar{T}_{22}\tilde{F}_1\right)\left(h_i(t)-h_j(t)\right)-\bar{T}_{21}\left(\dot{h}_i(t)-\dot{h}_j(t)\right)\right)=0. \tag{4.65}$$

Since $\hat{F}_1$ is of full row rank, one can find appropriate $v_i(t)$ ($i = 1, 2, \ldots, N$) to guarantee that (4.64) holds, for all $i \in \{1, 2, \ldots, N\}$ and $j \in N_i$. From (4.65), one sees that condition (i) is required. Since $\bar{J}$ consists of Jordan blocks corresponding to

λ_i ($i = 2, 3, \ldots, N$), to ensure that the system described by (4.57) is asymptotically stable, condition (ii) is required.

Sufficiency: For all $i \in \{1, 2, \ldots, N\}$ and $j \in N_i$, one can find $v_i(t) - v_j(t)$ satisfying (4.64). If condition (i) holds, from (4.64) and (4.65), one knows (4.63) holds. Since $\bar{T}$ is nonsingular, from (4.62) and (4.63), it can be obtained that for any $i \in \{1, 2, \ldots, N\}$ and $j \in N_i$, (4.61) holds. Then one has

$$\lim_{t\to\infty}\left(\begin{bmatrix} L \otimes (\bar{A}_{11} + \bar{B}_1 K_1) \\ L \otimes \left(\tilde{F}_1 + \tilde{B}_1 K_1\right) \end{bmatrix} h(t) - \begin{bmatrix} L \otimes I \\ 0 \end{bmatrix} \dot{h}(t) + \begin{bmatrix} L \otimes \bar{B}_1 \\ L \otimes \tilde{B}_1 \end{bmatrix} v(t)\right) = 0. \tag{4.66}$$

Substitute $L = UJU^{-1}$ into (4.66) and pre-multiply both sides of (4.66) by $\mathrm{diag}\{U^{-1} \otimes I, U^{-1} \otimes I\}$. It follows that

$$\lim_{t\to\infty}\left(\begin{bmatrix} \bar{J}\tilde{U} \otimes (\bar{A}_{11} + \bar{B}_1 K_1) \\ \bar{J}\tilde{U} \otimes \left(\tilde{F}_1 + \tilde{B}_1 K_1\right) \end{bmatrix} h(t) - \begin{bmatrix} \bar{J}\tilde{U} \otimes I \\ 0 \end{bmatrix} \dot{h}(t) + \begin{bmatrix} \bar{J}\tilde{U} \otimes \bar{B}_1 \\ \bar{J}\tilde{U} \otimes \tilde{B}_1 \end{bmatrix} v(t)\right) = 0. \tag{4.67}$$

Since the interaction topology has a spanning tree, by Lemma 2.1 and the structure of U, one knows that $\bar{J}$ is nonsingular. Pre-multiplying both sides of (4.67) by $\mathrm{diag}\{\bar{J}^{-1} \otimes I, \bar{J}^{-1} \otimes I\}$ yields

$$\lim_{t\to\infty}\left(\begin{bmatrix} \tilde{U} \otimes (\bar{A}_{11} + \bar{B}_1 K_1) \\ \tilde{U} \otimes \left(\tilde{F}_1 + \tilde{B}_1 K_1\right) \end{bmatrix} h(t) - \begin{bmatrix} \tilde{U} \otimes I \\ 0 \end{bmatrix} \dot{h}(t) + \begin{bmatrix} \tilde{U} \otimes \bar{B}_1 \\ \tilde{U} \otimes \tilde{B}_1 \end{bmatrix} v(t)\right) = 0. \tag{4.68}$$

By condition (ii), one knows that the system described by (4.57) is asymptotically stable. Therefore, from (4.57) and (4.68), it can be obtained that for any given bounded initial states, $\lim_{t\to\infty}\varsigma_y(t) = 0$. Then from Theorem 4.4, one sees that the time-varying output formation $h(t)$ can be achieved, hence it is feasible for swarm system (4.39). This completes the proof.

Remark 4.17 Theorem 4.6 reveals that the feasibility of the time-varying output formation $h(t)$ depends on the dynamics of each agent, the observable components of $(\bar{A}_{22}, \bar{A}_{12})$, external command $v(t)$, and the interaction topology. From (4.61), (4.64), and (4.65), one sees that by applying $v(t)$, the set of feasible output formation $h(t)$ can be expanded, and K_1 has no direct effect on the feasible set of $h(t)$ if $v(t)$ is applied.

Remark 4.18 If $K_1 = 0$, $K_2 = 0$, $h(t) \equiv 0$, $v(t) \equiv 0$ (or $C = I$, $K_2 = 0$, $h(t) \equiv 0$, $v(t) \equiv 0$), time-varying output formation problems in the current section become the output consensus problems discussed in [27] (or state consensus problems addressed in [21]). In addition, Theorem 3.6 in [27] and Theorem 1 in [21] are just special cases of Theorem 4.6.

From Theorem 4.6, the following corollary can be obtained directly.

Corollary 4.3 *If $v_i(t) \equiv 0$ $(i = 1, 2, \ldots, N)$, for any given bounded initial states, a time-varying output formation $h(t)$ is feasible for swarm system (4.39) if and only if condition (ii) in Theorem 4.6 holds, and for all $i \in \{1, 2, \ldots, N\}$,*

$$\lim_{t\to\infty}\left(\begin{bmatrix}\bar{A}_{11}+\bar{B}_1K_1\\ \tilde{F}_1+\tilde{B}_1K_1\end{bmatrix}\left(h_i(t)-h_j(t)\right)-\begin{bmatrix}I\\0\end{bmatrix}\left(\dot{h}_i(t)-\dot{h}_j(t)\right)\right)=0,\ j\in N_i. \tag{4.69}$$

Remark 4.19 From (4.69), one can see that K_1 can be used to expand the set of feasible time-varying output formation $h(t)$ when $v(t) \equiv 0$.

For simplicity of expression, let

$$A_o=\begin{bmatrix}\bar{A}_{11}+\bar{B}_1K_1+\bar{B}_1K_2 & \tilde{E}_1\\ \tilde{F}_1+\tilde{B}_1K_1+\tilde{B}_1K_2 & \tilde{D}_1\end{bmatrix},\ B_o=\begin{bmatrix}\bar{B}_1\\ \tilde{B}_1\end{bmatrix},\ C_o=\begin{bmatrix}I & 0\end{bmatrix},$$

then $\bar{\Upsilon}_i$ $(i = 2, 3, \ldots, N)$ in (4.54) can be rewritten as

$$\bar{\Upsilon}_i = A_o + \lambda_i B_o K_3 C_o\ (i = 2, 3, \ldots, N). \tag{4.70}$$

From (4.70), one sees that $\bar{\Upsilon}_i$ $(i = 2, 3, \ldots, N)$ are Hurwitz if and only if subsystems $(A_o, \lambda_i B_o, C_o)$ $(i = 2, 3, \ldots, N)$ can be stabilized by the static output feedback controller with gain matrix K_3 simultaneously. Given any real matrix R and any $\lambda \in \mathbb{C}$, let $\Lambda_R = \text{blockdiag}\{R, R\}$ and $\Psi_\lambda = \begin{bmatrix}\text{Re}(\lambda)I & -\text{Im}(\lambda)I\\ \text{Im}(\lambda)I & \text{Re}(\lambda)I\end{bmatrix}$. Based on the decomposition of the real and imaginary parts, the condition that $\bar{\Upsilon}_i$ $(i = 2, 3, \ldots, N)$ are Hurwitz is equivalent to that subsystems $(\Lambda_{A_o}, \Psi_{\lambda_i}\Lambda_{B_o}, \Lambda_{C_o})$ $(i = 2, 3, \ldots, N)$ can be stabilized by the static output feedback controller with gain matrix Λ_{K_3} simultaneously. Define $\tilde{\lambda}_{1,2} = \text{Re}(\lambda_2) \pm j\mu_{\bar{N}}$ and $\tilde{\lambda}_{3,4} = \text{Re}(\lambda_N) \pm j\mu_{\bar{N}}$, where $j^2 = -1$ and $\mu_{\bar{N}} = \max\{\text{Im}(\lambda_i), i = 2, 3, \ldots, N\}$. Let Ω_0, Ω_1 and Ω_2 be real symmetric matrices independent of λ_i $(i = 2, 3, \ldots, N)$ and $\tilde{\lambda}_i$ $(i = 1, 2, 3, 4)$. The following lemma holds.

Lemma 4.3 ([24]) *If $\Omega_0 + \text{Re}\left(\tilde{\lambda}_i\right)\Omega_1 + \text{Im}\left(\tilde{\lambda}_i\right)\Omega_2 < 0$ $(i = 1, 2, 3, 4)$, then $\Omega_0 + \text{Re}(\lambda_i)\Omega_1 + \text{Im}(\lambda_i)\Omega_2 < 0$ $(i = 2, 3, \ldots, N)$.*

Theorem 4.7 *For any given bounded initial states, swarm system (4.36) achieves time-varying output formation $h(t)$ by protocol (4.38) if condition (i) in Theorem 4.6 holds, and there exist real matrices K_3 and $R_o = R_o^T > 0$ such that*

$$(\Lambda_{A_o} + \Psi_{\tilde{\lambda}_i}\Lambda_{B_o}\Lambda_{K_3}\Lambda_{C_o})^T R_o + R_o(\Lambda_{A_o} + \Psi_{\tilde{\lambda}_i}\Lambda_{B_o}\Lambda_{K_3}\Lambda_{C_o}) < 0\ (i = 1, 2, 3, 4). \tag{4.71}$$

Proof Let $\Pi_i = (\Lambda_{A_o} + \Psi_{\lambda_i}\Lambda_{B_o}\Lambda_{K_3}\Lambda_{C_o})^T R_o + R_o(\Lambda_{A_o} + \Psi_{\lambda_i}\Lambda_{B_o}\Lambda_{K_3}\Lambda_{C_o})$ $(i = 2, 3, \ldots, N)$, then there exist Ω_0, Ω_1, and Ω_2 such that Π_i can be rewritten as

$$\Pi_i = \Omega_0 + \mathrm{Re}(\lambda_i)\Omega_1 + \mathrm{Im}(\lambda_i)\Omega_2. \tag{4.72}$$

Similarly, let $\tilde{\Pi}_i = (\Lambda_{A_o} + \Psi_{\tilde{\lambda}_i}\Lambda_{B_o}\Lambda_{K_3}\Lambda_{C_o})^T R_o + R_o(\Lambda_{A_o} + \Psi_{\tilde{\lambda}_i}\Lambda_{B_o}\Lambda_{K_3}\Lambda_{C_o})$ $(i = 1, 2, 3, 4)$, then one has

$$\tilde{\Pi}_i = \Omega_0 + \mathrm{Re}\left(\tilde{\lambda}_i\right)\Omega_1 + \mathrm{Im}\left(\tilde{\lambda}_i\right)\Omega_2. \tag{4.73}$$

By (4.72), (4.73), and Lemma 4.3, it can be obtained that if condition (4.71) holds, then $\Pi_i < 0$ $(i = 2, 3, \ldots, N)$.

Consider the stability of the following subsystem:

$$\dot{\varphi}_i(t) = \left(\Lambda_{A_o} + \Psi_{\lambda_i}\Lambda_{B_o}\Lambda_{K_3}\Lambda_{C_o}\right)\varphi_i(t), \tag{4.74}$$

where $i \in \{2, 3, \ldots, N\}$. Construct the following Lyapunov functional candidate:

$$V_i(t) = \varphi_i^T(t)R_o\varphi_i(t). \tag{4.75}$$

Taking the time derivative of $V_i(t)$ along the trajectory of (4.74), one has

$$\begin{aligned}\dot{V}_i(t) = &\ \varphi_i^T(t)(\Lambda_{A_o} + \Psi_{\lambda_i}\Lambda_{B_o}\Lambda_{K_3}\Lambda_{C_o})^T R_o\varphi_i(t)\\ &+ \varphi_i^T(t)R_o(\Lambda_{A_o} + \Psi_{\lambda_i}\Lambda_{B_o}\Lambda_{K_3}\Lambda_{C_o})\varphi_i(t).\end{aligned}$$

Since $\Pi_i < 0$ $(i = 2, 3, \ldots, N)$, it can be verified that $\dot{V}_i(t) < 0$, which means that $\bar{\Upsilon}_i$ $(i = 2, 3, \ldots, N)$ are Hurwitz. From Theorem 4.6, the conclusion of Theorem 4.7 can be obtained.

Remark 4.20 Theorem 4.7 decreases the calculation complexity by reducing the simultaneous stabilization problems of $N - 1$ subsystems in Theorem 4.6 into those of four ones. Since one of the main features of swarm systems is of large scale, thus the solvability is important. Note that the condition (4.71) in Theorem 4.7 is independent of the number of agents. As a result, for high-order LTI swarm systems with directed interaction topologies and a huge number of agents, the calculation complexity can be decreased significantly, although it may result in some conservation.

Using the PBH criterion for stabilizability, it can be shown that if (A, B) is stabilizable, then $(\Lambda_{A_o}, \Psi_{\tilde{\lambda}_i}\Lambda_{B_o})$ $(i = 1, 2, 3, 4)$ are stabilizable. Since $(\tilde{D}_1, \tilde{E}_1)$ is completely observable, according to the PBH criterion for observability, one can obtain that $(\Lambda_{C_o}, \Lambda_{A_o})$ is detectable. Then the algorithm proposed by He and Wang [28] can be applied to solve the quadratic matrix inequalities (QMIs) in Theorem 4.7 and the convergence of it can be guaranteed. Algorithm 4.2 consists of a cone complementarity linearization (CCL) algorithm and an interactive linear matrix inequality

(ILMI) algorithm, where the CCL algorithm is used to find an initial R_o and the ILMI algorithm is applied to determine K_3 in protocol (4.38).

Algorithm 4.2.
Part I (CCL):
Step 1: Set $k = 1$, $R_0 = I$, and $Q_0 = I$, and give tolerances $\chi_1 > 0$ and $\chi_2 > 0$.
Step 2: Solve the following optimization problem for R_k, Q_k, M_1, and M_2:
OP 1:

$$\begin{aligned} &\text{minimize} \quad \text{trace}(R_k Q_{k-1} + Q_k R_{k-1}) \\ &\text{subject to } \Lambda_{A_o} R_k + R_k \Lambda_{A_o}^T + \Psi_{\tilde{\lambda}_i} \Lambda_{B_o} M_1 + M_1^T \Lambda_{B_o}^T \Psi_{\tilde{\lambda}_i}^T < 0 \ (i = 1, 2, 3, 4), \\ &Q_k \Lambda_{A_o} + \Lambda_{A_o}^T Q_k + M_2 \Lambda_{C_o} + \Lambda_{C_o}^T M_2^T < 0, \\ &\begin{bmatrix} R_k & I \\ I & Q_k \end{bmatrix} \geqslant 0. \end{aligned}$$

Step 3: If $\text{trace}(R_k Q_k) - 2(q + g) < \chi_1$, then let $R = R_k$ and go to Part II.
Step 4: If $\text{trace}(R_k Q_k) - \text{trace}(R_{k-1} Q_{k-1}) < \chi_2$, then go to Step 6.
Step 5: Set $k = k + 1$, go to Step 2.
Step 6: Swarm system (4.36) may not be able to achieve the time-varying output formation $h(t)$ by protocol (4.38). Stop.
Part II (ILMI):
Step 1: Set $k = 1$ and $R_1 = R$, and give a tolerance $\chi_3 > 0$.
Step 2: Solve the following optimization problem for K_3 with a given R_k:
OP 2:

$$\text{minimize} \quad \mu_k$$

$$\text{subject to } \left(\Lambda_{A_o} + \Psi_{\tilde{\lambda}_i} \Lambda_{B_o} \Lambda_{K_3} \Lambda_{C_o}\right) R_k + R_k \left(\Lambda_{A_o} + \Psi_{\tilde{\lambda}_i} \Lambda_{B_o} \Lambda_{K_3} \Lambda_{C_o}\right)^T - \mu_k R_k < 0. \tag{4.76}$$

Step 3: If $\mu_k \leqslant 0$, K_3 is found. Stop.
Step 4: Set $k = k + 1$, and solve the following optimization problem for R_k with a given K_3:
OP 3:

$$\begin{aligned} &\text{minimize} \quad \mu_k \\ &\text{subject to (4.76)} \end{aligned}$$

Step 5: If $\mu_k \leqslant 0$, K_3 is found. Stop.
Step 6: Solve the following optimization problem for R_k with given K_3 and μ_k:
OP 4:

$$\begin{aligned} &\text{minimize} \quad \text{trace}(R_k) \\ &\text{subject to (4.76)} \end{aligned}$$

Step 7: If $\|R_k - R_{k-1}\|/\|R_k\| < \chi_3$, go to Step 8, else let $k = k + 1$ and $R_k = R_{k-1}$ and go to Step 2.
Step 8: Swarm system (4.36) may not be able to achieve the time-varying output formation $h(t)$ by protocol (4.38). Stop.

Based on the above results, an algorithm to design protocol (4.38) for swarm system (4.36) to achieve time-varying output formation $h(t)$ can be summarized as follows:

Algorithm 4.3.
Step 1: Check the feasible condition (i) in Theorem 4.6. If it is satisfied, then $v_j(t)$ ($j = 1, 2, \ldots, N$) can be determined by Eq. (4.64), and K_1 can be any constant matrix with appropriate dimension, e.g., $K_1 = 0$; else stop.

If it is required that $v(t) \equiv 0$, solve feasible condition (4.69) for K_1. If there exists constant gain matrix K_1 satisfying condition (4.69), then continue; else stop.
Step 2: Choose K_2 to partially design the motion modes of the time-varying output formation reference by partially assigning the eigenvalues of $A + BK_1C + BK_2C$ at the desired locations in the complex plane.
Step 3: Design K_3 to make system (4.57) asymptotically stable using Algorithm 4.2.

Remark 4.21 It should be pointed out that by Eq. (4.64), $v_i(t)$ ($i = 1, 2, \ldots, N$) cannot be uniquely determined. One can first specify a $v_i(t)$ ($i \in \{1, 2, \ldots, N\}$), then determine the other $v_j(t)$ ($j \in \{1, 2, \ldots, N\}$, $j \neq i$) by Eq. (4.64). Since $\hat{F}_1$ is of full row rank, the existence of $v_i(t)$ can be guaranteed.

4.3.4 Numerical Simulations

In this subsection, a numerical example is given to illustrate the effectiveness of theoretical results obtained in this section.

Example 4.3 Consider a sixth-order swarm system with eight agents. The interaction topology of the swarm system is shown in Fig. 4.7.

The dynamics of each agent is described by (4.36) with $x_i(t) = [x_{i1}(t), x_{i2}(t), \ldots, x_{i6}(t)]^T$, $y_i(t) = [y_{i1}(t), y_{i2}(t), y_{i3}(t)]^T$ ($i = 1, 2, \ldots, 8$), and

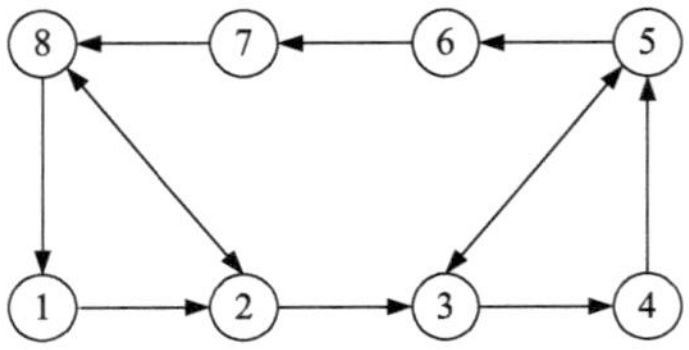

Fig. 4.7 Directed interaction topology G

$$A = \begin{bmatrix} 10 & 10 & 28 & -2.5 & 16.5 & -7 \\ 10 & 9 & 28 & -5.5 & 13.5 & -3 \\ -10 & -9 & -27 & 3.5 & -14.5 & 5 \\ -4 & -8 & -14 & -1.5 & -5.5 & 3 \\ 2 & 4 & 10 & -3.5 & 2.5 & -1 \\ -5 & -6 & -16 & 4 & -6 & 0 \end{bmatrix}, \; B = \begin{bmatrix} -1 \\ -1 \\ 1 \\ 1 \\ 1 \\ 0 \end{bmatrix},$$

$$C = \begin{bmatrix} 1 & 0 & 1 & 0 & 0 & 0 \\ 0 & 1 & 1 & 0 & 0 & 0 \\ 0 & 0 & 1 & -1 & 1 & 0 \end{bmatrix}.$$

Choose

$$\bar{C} = \begin{bmatrix} 0 & 0 & 0 & 0 & 0 & 1 \\ 0 & 0 & 0 & 1 & 0 & 0 \\ 0 & 0 & 0 & 0 & 1 & 0 \end{bmatrix}.$$

It can be shown that (A, B) is stabilizable and $(\bar{A}_{22}, \bar{A}_{12})$ is not completely observable. Choose a nonsingular matrix $\tilde{T}$ as follows:

$$\tilde{T} = \begin{bmatrix} 1 & 0 & 1 \\ 0 & -1 & 2 \\ 0 & 1 & 0 \end{bmatrix}.$$

It can be obtained that

$$\bar{T} = \begin{bmatrix} 0 & 0 & 0 & 0 & 1 \\ 0 & 1 & 0 & 0 & 0 \\ 0 & 0 & 1 & 0 & -1 \\ 0 & 0 & 0 & 1 & 1 \\ 1 & 0 & 0 & 0 & 0 \end{bmatrix}, \; \hat{T}^{-1} = I.$$

The outputs of these eight agents are required to preserve a periodic time-varying parallel octagon formation and keep rotating around the predefined time-varying output formation reference. The output formation is defined as follows:

$$h_i(t) = \begin{bmatrix} 10 \sin\left(t + \frac{(i-1)\pi}{4}\right) \\ 10 \cos\left(t + \frac{(i-1)\pi}{4}\right) \\ -10 \sin\left(t + \frac{(i-1)\pi}{4}\right) \end{bmatrix} \quad (i = 1, 2, \ldots, 8).$$

If the time-varying output formation specified by the above $h_i(t)$ $(i = 1, 2, \ldots, 8)$ is achieved, the outputs of eight agents will locate at the eight vertices of a parallel octagon, respectively, and keep rotating with angular velocity 1rad/s. In addition, the edge length of the desired parallel octagon is periodic time-varying.

It can be verified that condition (i) in Theorem 4.6 is satisfied. According to Algorithm 4.3, K_1 can be chosen as $K_1 = [0, 0, 0]$ and $v_i(t)$ $(i = 1, 2, \ldots, 8)$ can be obtained as

$$v_i(t) = 20\sin\left(t + \frac{\pi}{3}(i-1)\right) - 40\cos\left(t + \frac{\pi}{3}(i-1)\right) \ (i = 1, 2, \ldots, 8).$$

Choose $K_2 = [-6, -2, -5.3]$ to assign the eigenvalues of $A + BK_1C + BK_2C$ at -8.2498, $-0.6215 + 3.8875j$, $-0.6215 - 3.8875j$, -2, -0.8261 and 0.0189 with $j^2 = -1$. In this case, the output formation reference will diverge with a slow speed. By Algorithm 4.2, K_3 can be obtained to make system (4.57) asymptotically stable as follows: $K_3 = [0.4227, -0.0363, 0.1183]$.

For simplicity of description, let the initial states of each agent be $x_{ij}(0) = i\,(\Theta - 0.5)$, $(i = 1, 2, \ldots, 8;\ j = 1, 2, \ldots, 6)$. Figure 4.8 shows the output snapshots of the eight agents and the snapshots of predefined time-varying output formation reference function at different time, where the outputs of agents are denoted by the point, circle, x-mark, asterisk, square, diamond, triangle, and plus, and those of the predefined output formation reference function are represented by the pentagram, respectively. Figure 4.8a, b indicates that the outputs of the swarm system achieve a parallel octagon formation and the point corresponding to the predefined formation reference function $r(t)$ lies in the center of the formation. Figure 4.8b–d shows that

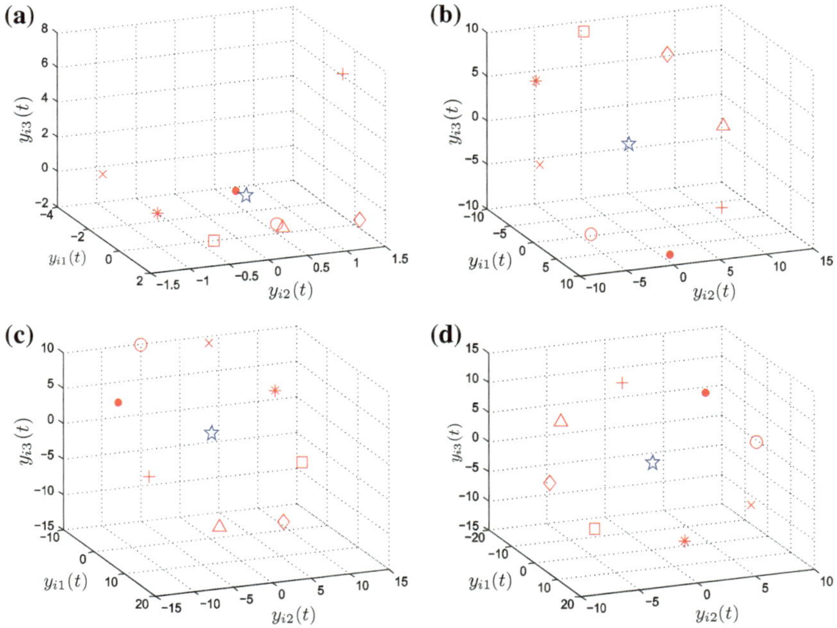

Fig. 4.8 Output snapshots of eight agents and the formation reference. **a** $t = 0$ s. **b** $t = 146$ s. **c** $t = 148$ s. **d** $t = 150$ s

the achieved output formation keeps rotation around the predefined output formation reference, and both parallel octagon formation and the output formation reference are time-varying. Therefore, the desired time-varying output formation is achieved.

4.4 Time-Varying Formation Control and Experiments for UAV Swarm Systems

In this subsection, time-varying formation control analysis and design problems for UAV swarm systems are investigated. To achieve predefined time-varying formations, formation protocols are presented for UAV swarm systems first, where the velocities of UAVs can be different when achieving formations. Then consensus-based approaches are applied to deal with the time-varying formation control problems for UAV swarm systems. Necessary and sufficient conditions for UAV swarm systems to achieve time-varying formations are proposed. An explicit expression of the time-varying formation reference function is derived. Moreover, a procedure to design the protocol for UAV swarm systems to achieve time-varying formations is given. It is shown that choosing appropriate initial states for each UAV, the collision avoidance among UAVs can be guaranteed. Finally, a quadrotor formation platform which consists of five quadrotors is introduced. Theoretical results obtained in this section are validated on the quardrotor formation platform, and outdoor experimental results are presented.

4.4.1 Problem Description

Consider a UAV swarm system with N UAVs. For each of these UAVs, since the trajectory dynamics has much larger time constants than the attitude dynamics, if the formation is only concerned with positions and velocities, then the formation control can be implemented with an inner/outer-loop structure [29, 30]. In this configuration, the outer-loop can be used to drive the UAV towards the desired position with desired velocity while the inner-loop can be used to track the attitude. The schematic diagram of the two-loop configuration for formation control is depicted in Fig. 4.9. The current paper mainly focuses on designing the outer-loop. Therefore, on the formation control level, a UAV can be regarded as a point-mass system, and the dynamics of each UAV can be approximately described by the following double integrator [3, 31–33]:

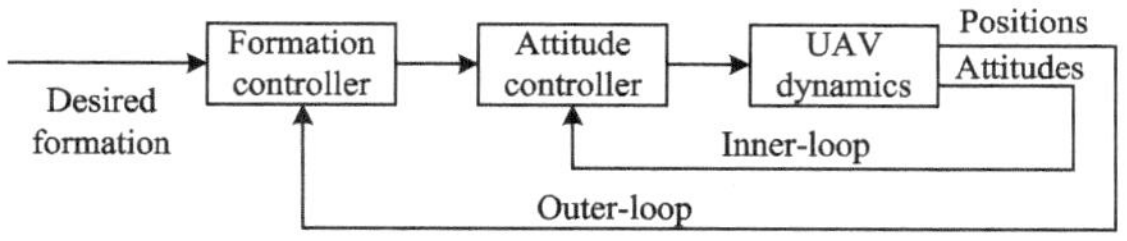

Fig. 4.9 Schematic diagram of the two-loop configuration for formation control

$$\begin{cases} \dot{x}_i(t) = v_i(t), \\ \dot{v}_i(t) = u_i(t), \end{cases} \tag{4.77}$$

where $i=1, 2, \ldots, N$, $x_i(t) \in \mathbb{R}^n$ and $v_i(t) \in \mathbb{R}^n$ denote the position and velocity vectors of UAV i, respectively, and $u_i(t) \in \mathbb{R}^n$ are the control inputs. In the following, for simplicity of description, it is assumed that $n = 1$, if not otherwise specified. However, it should be pointed out that similar analysis can also be done for the higher dimensional case by using Kronecker product, and all the results hereafter remain valid for $n > 1$.

Let $\zeta_i(t) = [x_i(t), v_i(t)]^T$, $B_1 = [1, 0]^T$, and $B_2 = [0, 1]^T$. Then UAV swarm system (4.77) can be rewritten as

$$\dot{\zeta}_i(t) = B_1 B_2^T \zeta_i(t) + B_2 u_i(t). \tag{4.78}$$

Let $h_i(t) = [h_{ix}(t), h_{iv}(t)]^T$ $(i = 1, 2, \ldots, N)$ be piecewise continuously differentiable vectors and $h(t) = [h_1^T(t), h_2^T(t), \ldots, h_N^T(t)]^T \in \mathbb{R}^{2N}$. Consider the following time-varying formations protocol:

$$u_i(t) = K_1 (\zeta_i(t) - h_i(t)) + K_2 \sum_{j \in N_i} w_{ij} \left((\zeta_j(t) - h_j(t)) - (\zeta_i(t) - h_i(t))\right) + \dot{h}_{iv}(t), \tag{4.79}$$

where $i = 1, 2, \ldots, N$, $K_1 = [k_{11}, k_{12}]$ and $K_2 = [k_{21}, k_{22}]$.

Let $\zeta(t) = [\zeta_1^T(t), \zeta_2^T(t), \ldots, \zeta_N^T(t)]^T$, $h_x(t) = [h_{1x}(t), h_{2x}(t), \ldots, h_{Nx}(t)]^T$ and $h_v(t) = [h_{1v}(t), h_{2v}(t), \ldots, h_{Nv}(t)]^T$. Under protocol (4.79), UAV swarm system (4.78) can be written in a compact form as follows:

$$\begin{aligned} \dot{\zeta}(t) = & \left(I_N \otimes \left(B_2 K_1 + B_1 B_2^T\right) - L \otimes (B_2 K_2)\right) \zeta(t) + (I_N \otimes B_2)\, \dot{h}_v(t) \\ & - (I_N \otimes (B_2 K_1) - L \otimes (B_2 K_2))\, h(t). \end{aligned} \tag{4.80}$$

The present section mainly investigates the following three problems for UAV swarm system (4.80): (i) under what conditions the time-varying formations $h(t)$ can be achieved; (ii) how to design protocol (4.79) to achieve the time-varying formations $h(t)$; and (iii) how to demonstrate the theoretical results on practical quadrotor formation platform.

4.4.2 Time-Varying Formation Analysis and Protocol Design

In this section, first, time-varying formation problems for UAV swarm system (4.80) are transformed into consensus problems. Then, necessary and sufficient conditions for UAV swarm system (4.80) to achieve time-varying formations $h(t)$ are presented, and an explicit expression of the time-varying formation reference function is given.

Second, a procedure to determine the gain matrices in protocol (4.79) is proposed. Finally, an approach to avoid collision is given.

Construct the similar nonsingular transformation matrices U and U^{-1} as the ones in Sect. 4.2, such that $U^{-1}LU = J$. Let $\tilde{\zeta}_i(t) = \zeta_i(t) - h_i(t)$ $(i = 1, 2, \ldots, N)$. Then swarm system (4.80) can be rewritten as follows:

$$\dot{\tilde{\zeta}}(t) = \left(I_N \otimes \left(B_2K_1 + B_1B_2^T\right) - L \otimes (B_2K_2)\right)\tilde{\zeta}(t) + (I_N \otimes B_1)(h_v(t) - \dot{h}_x(t)). \tag{4.81}$$

The following lemma holds directly.

Lemma 4.4 *UAV swarm system (4.80) achieves time-varying formation $h(t)$ if and only if swarm system (4.81) achieves consensus.*

Define the ν in Sect. 2.2 as $\nu = 2$. Let $\theta_U(t) = (\tilde{u}_1^H \otimes I_2)\tilde{\zeta}(t)$ and $\varsigma_U(t) = (\tilde{U} \otimes I_2)\tilde{\zeta}(t)$. Then swarm system (4.81) can be transformed into

$$\dot{\theta}_U(t) = \left(B_2K_1 + B_1B_2^T\right)\theta_U(t) + \left(\tilde{u}_1^H \otimes B_1\right)(h_v(t) - \dot{h}_x(t)), \tag{4.82}$$

$$\begin{aligned}\dot{\varsigma}_U(t) = &\left(I_{N-1} \otimes \left(B_2K_1 + B_1B_2^T\right) - \bar{J} \otimes (B_2K_2)\right)\varsigma_U(t) \\ &+ \left(\tilde{U} \otimes B_1\right)(h_v(t) - \dot{h}_x(t)).\end{aligned} \tag{4.83}$$

Lemma 4.5 ([34]) *Let A_U be a 2×2 dimensional complex matrix with characteristic polynomial $f(s) = s^2 + a_1s + a_2$, where s is a complex variable. The system $\dot{\varphi}(t) = A_U\varphi(t)$ is asymptotically stable if and only if* $\mathrm{Re}(a_1) > 0$ *and* $\mathrm{Re}(a_1)\mathrm{Re}(a_1\bar{a}_2) - \mathrm{Im}(a_2)^2 > 0$.

The following theorem presents a necessary and sufficient condition for UAV swarm system (4.80) to achieve time-varying formation $h(t)$.

Theorem 4.8 *For any given bounded initial states, UAV swarm system (4.80) achieves time-varying formation $h(t)$ if and only if the following conditions hold simultaneously:*
(i) For any $i \in \{1, 2, \ldots, N\}$,

$$\lim_{t\to\infty}\left((h_{iv}(t) - h_{jv}(t)) - (\dot{h}_{ix}(t) - \dot{h}_{jx}(t))\right) = 0,\ j \in N_i; \tag{4.84}$$

(ii) For any $i \in \{2, 3, \ldots, N\}$,

$$-k_{12} + \mathrm{Re}(\lambda_i)k_{22} > 0, \tag{4.85}$$

$$(-k_{12} + \mathrm{Re}(\lambda_i)k_{22})\,\psi_i - \mathrm{Im}(\lambda_i)^2 k_{21}^2 > 0, \tag{4.86}$$

where

$$\psi_i = k_{12}k_{11} - \mathrm{Re}(\lambda_i)\,(k_{12}k_{21} + k_{11}k_{22}) + \left(\mathrm{Re}(\lambda_i)^2 + \mathrm{Im}(\lambda_i)^2\right) k_{21}k_{22}.$$

Proof Let $\tilde{\zeta}_C(t) = (U \otimes I_2)[\theta_U^H(t), 0]^H$ and $\tilde{\zeta}_{\bar{C}}(t) = (U \otimes I_2)[0, \varsigma_U^H(t)]^H$. Because c_1 and c_2 are linearly independent vectors, there exist $\alpha_1(t)$, $\alpha_2(t)$ and $\alpha_{2i+k}(t)$ $(i = 1, 2, \ldots, N-1;\ k = 1, 2)$ such that $\theta_U(t) = \alpha_1(t)c_1 + \alpha_2(t)c_2$ and $\varsigma_U(t) = \left[\alpha_3(t)c_1^H + \alpha_4(t)c_2^H, \ldots, \alpha_{2N-1}(t)c_1^H + \alpha_{2N}(t)c_2^H\right]^H$. Due to $[\theta_U^H(t), 0]^H = e_1 \otimes \theta_U(t)$, one has

$$\tilde{\zeta}_C(t) = (U \otimes I_2)\,(e_1 \otimes \theta_U(t)) = \bar{u}_1 \otimes \theta_U(t) = \alpha_1(t)p_1 + \alpha_2(t)p_2 \in \mathbb{C}(U). \tag{4.87}$$

By the structures of p_j $(j = 3, 4, \ldots, 2N)$, it can be shown

$$\tilde{\zeta}_{\bar{C}}(t) = \sum_{i=2}^{N} (\alpha_{2i-1}(t)\,(\bar{u}_i \otimes c_1) + \alpha_{2i}(t)\,(\bar{u}_i \otimes c_2)) = \sum_{j=3}^{2N} \alpha_j(t)p_j \in \overline{\mathbb{C}}(U). \tag{4.88}$$

Note that $[\theta_U^H(t), \varsigma_U^H(t)]^H = (U^{-1} \otimes I_2)\tilde{\zeta}(t)$, one has $\tilde{\zeta}(t) = \tilde{\zeta}_C(t) + \tilde{\zeta}_{\bar{C}}(t)$. From Lemmas 4.4 and 2.3, for any given bounded initial states, UAV swarm system (4.80) achieves time-varying formation $h(t)$ if and only if $\lim_{t\to\infty}\tilde{\zeta}_{\bar{C}}(t) = 0$; that is,

$$\lim_{t\to\infty} \varsigma_U(t) = 0. \tag{4.89}$$

Let

$$\dot{\tilde{\varsigma}}_U(t) = \left(I_{N-1} \otimes \left(B_2K_1 + B_1B_2^T\right) - \bar{J} \otimes (B_2K_2)\right) \tilde{\varsigma}_U(t). \tag{4.90}$$

From (4.83) and (4.89), one knows that UAV swarm system (4.80) achieves formation $h(t)$ if and only if the system described by (4.90) is asymptotically stable and

$$\lim_{t\to\infty} \left(\tilde{U} \otimes B_1\right) \left(h_v(t) - \dot{h}_x(t)\right) = 0. \tag{4.91}$$

Next, it will be shown that conditions (i) and (ii) in Theorem 4.8 are equivalent to the conditions that (4.91) holds and system (4.90) is asymptotically stable, respectively.

If (4.84) holds, one has

$$\lim_{t\to\infty} (L \otimes B_1)\left(h_v(t) - \dot{h}_x(t)\right) = 0. \tag{4.92}$$

Substituting $L = UJU^{-1}$ into (4.92) and pre-multiplying both sides of (4.92) by $U^{-1} \otimes I$, one has

$$\lim_{t\to\infty}\left(\bar{J}\tilde{U}\otimes B_1\right)\left(h_v(t)-\dot{h}_x(t)\right)=0. \tag{4.93}$$

Note that G has a spanning tree, one gets that $\bar{J}$ is nonsingular. Pre-multiplying both sides of (4.93) by $\bar{J}^{-1} \otimes I_2$ results in (4.91); that is, condition (i) is sufficient for (4.91).

If (4.91) holds, let $\tilde{U} = [\hat{U}, \hat{u}]$, where $\hat{U} \in \mathbb{C}^{(N-1)\times(N-1)}$ and $\hat{u} \in \mathbb{C}^{(N-1)\times 1}$ with $\hat{u}$ being the last column vector of $\tilde{U}$. Since $\text{rank}(\tilde{U}) = N-1$, without loss of generality, it is assumed that $\text{rank}(\hat{U}) = N-1$. From (4.91), one has

$$\lim_{t\to\infty}\left([\hat{U},\hat{u}]\otimes B_1\right)\left(h_v(t)-\dot{h}_x(t)\right)=0. \tag{4.94}$$

Since $\tilde{U}\mathbf{1}_N = 0$, it follows that

$$\hat{u} = -\hat{U}\mathbf{1}_{N-1}. \tag{4.95}$$

Let $\bar{h}_x(t) = [h_{1x}(t), h_{2x}(t), \ldots, h_{(N-1)x}(t)]^T$ and $\bar{h}_v(t) = [\dot{h}_{1v}(t), \dot{h}_{2v}(t), \ldots, \dot{h}_{(N-1)v}(t)]^T$. From (4.94) and (4.95), it can be verified that

$$\begin{aligned}\lim_{t\to\infty}\left(\hat{U}\otimes I_2\right)\Big(&(I_{N-1}\otimes B_1)\left(\bar{h}_v(t)-\dot{\bar{h}}_x(t)\right)\\ &-(\mathbf{1}_{N-1}\otimes B_1)\left(h_{Nv}(t)-\dot{h}_{Nx}(t)\right)\Big)=0.\end{aligned} \tag{4.96}$$

Pre-multiplying the both sides of (4.96) by $\hat{U}^{-1} \otimes I_2$, one has for all $i \in \{1, 2, \ldots, N-1\}$

$$\lim_{t\to\infty}\left((h_{iv}(t)-h_{Nv}(t))-\left(\dot{h}_{ix}(t)-\dot{h}_{Nx}(t)\right)\right)=0. \tag{4.97}$$

By (4.97), it can be shown that condition (i) holds. Therefore, condition (i) in Theorem 4.8 is equivalent to condition (4.91).

From the structure of $\bar{J}$, one knows that the stability of system (4.90) is equivalent to these of the following $N-1$ subsystems

$$\dot{\varsigma}_{Ui}(t)=\left(B_2\left(K_1-\lambda_i K_2\right)+B_1B_2^T\right)\varsigma_{Ui}(t)\ (i=2,3,\ldots,N). \tag{4.98}$$

It can be obtained that the characteristic polynomials of subsystems (4.98) are $f_i(s) = s^2 - (k_{12} - \lambda_i k_{22})\,s - (k_{11} - \lambda_i k_{21})$ $(i = 2, 3, \ldots, N)$. From Lemma 4.5, one gets that system (4.90) is asymptotically stable if and only if condition (ii) holds. This completes the proof for Theorem 4.8.

Remark 4.22 Formation control problems for swarm system (4.78) have been investigated in [11] and [14], and sufficient conditions have been obtained to achieve time invariant formations. By Theorem 4.8, necessary and sufficient conditions can be obtained for swarm system (4.78) to achieve time-varying formations. Moreover, by Theorem 4.8, it can be verified that the conditions in [14] are not only sufficient but also necessary.

If UAV swarm system (4.80) achieves time-varying formations $h(t)$, then from the proof of Theorem 4.8, one knows that $\tilde{\zeta}_i(t) - \theta_U(t) \rightarrow 0$ as $t \rightarrow \infty$. Therefore, subsystem (4.82) determines the formation reference function and an explicit expression of the time-varying formation reference function can be obtained directly as follows:

Lemma 4.6 *If UAV swarm system (4.80) achieves time-varying formations $h(t)$ with the time-varying formation reference function $r(t)$, then*

$$\lim_{t\rightarrow\infty}(r(t) - r_0(t) - r_h(t)) = 0,$$

where

$$r_0(t) = e^{(B_2K_1+B_1B_2^T)t}\left(\tilde{u}_1^H \otimes I_2\right)\zeta(0),$$

$$r_h(t) = \int_0^t e^{(B_2K_1+B_1B_2^T)(t-\tau)}\left(\tilde{u}_1^H \otimes B_2\right)\left(\dot{h}_v(\tau) - k_{12}h_v(\tau) - k_{11}h_x(\tau)\right)d\tau - \left(\tilde{u}_1^H \otimes I_2\right)h(t).$$

In the following, a procedure to determine the gain matrices in protocol (4.79) for UAV swarm system (4.78) to achieve time-varying formations is proposed.

Theorem 4.9 *If condition (i) in Theorem 4.8 holds, for any given bounded initial states, UAV swarm system (4.78) achieves time-varying formations by protocol (4.79) with $K_2 = [\mathrm{Re}(\lambda_2)]^{-1}B_2^TP_U$ where P_U is the positive definite solution to the algebraic Riccati equation*

$$P_U\left(B_2K_1 + B_1B_2^T\right) + \left(B_2K_1 + B_1B_2^T\right)^TP_U - P_UB_2B_2^TP_U + I = 0. \tag{4.99}$$

Proof Consider the $N-1$ subsystems described by (4.98) and construct the following Lyapunov function candidate

$$V_i(t) = \bar{\varsigma}_{Ui}^H(t)P_U\bar{\varsigma}_{Ui}(t)\ (i = 2, 3, \ldots, N). \tag{4.100}$$

Let $K_2 = [\mathrm{Re}(\lambda_2)]^{-1}B_2^TP_U$. Taking the derivative of $V_i(t)$ with respect to t along the solution of subsystem (4.98), one has

$$\dot{V}_i(t) = -\bar{\varsigma}_{Ui}^H(t)\bar{\varsigma}_{Ui}(t) + \left(1 - 2\mathrm{Re}(\lambda_i)[\mathrm{Re}(\lambda_2)]^{-1}\right)\bar{\varsigma}_{Ui}^H(t)P_U B_2 B_2^T \bar{P}_U \bar{\varsigma}_{Ui}(t). \quad (4.101)$$

Since $0 < \mathrm{Re}(\lambda_2) \leq \cdots \leq \mathrm{Re}(\lambda_N)$, from (4.101), it holds that $\dot{V}_i(t) \leqslant -\bar{\varsigma}_{Ui}^H(t)\bar{\varsigma}_{Ui}(t)(i = 2, 3, \ldots, N)$. According to Lyapunov's second method for stability, the $N - 1$ subsystems described by (4.98) are asymptotically stable. From the proof of Theorem 4.8, one knows that swarm system (4.78) achieves time-varying formations by protocol (4.79). The proof for Theorem 4.9 is completed.

Based on the above results, a design procedure of protocol (4.79) can be summarized as follows. First, choose K_1 to design the motion modes of the formation reference by assigning the eigenvalues of $B_2K_1 + B_1B_2^T$ at the desired locations in the complex plane. Since $(B_1B_2^T, B_2)$ is controllable, there always exists such K_1. Second, design K_2 to satisfy condition (4.86) using the conclusion of Theorem 4.9.

Next, it will be shown that by choosing appropriate formation and initial states for each agent, collision avoidance can be ensured.

Let $U = [\bar{u}_1, \bar{U}]$. Since $\bar{u}_1 = \mathbf{1}_N$ and $\tilde{\xi}(t) = (U \otimes I)[\theta_U^H(t), \varsigma_U^H(t)]^H$, one gets

$$\tilde{\xi}(t) = (\mathbf{1}_N \otimes I)\theta_U(t) + (\bar{U} \otimes I)\varsigma_U(t). \quad (4.102)$$

Choose $h_v(t)$ and $\dot{h}_x(t)$ such that

$$\left(\tilde{U} \otimes B_1\right)(h_v(t) - \dot{h}_x(t)) = 0. \quad (4.103)$$

From (4.83) and (4.103), one has

$$\varsigma_U(t) = e^{A_\varsigma t}\varsigma_U(0) = e^{A_\varsigma t}(\tilde{U} \otimes I)\tilde{\xi}(0), \quad (4.104)$$

where

$$A_\varsigma = I_{N-1} \otimes \left(B_2K_1 + B_1B_2^T\right) - \bar{J} \otimes (B_2K_2).$$

Then Eq. (4.102) can be rewritten as

$$\tilde{\xi}(t) = (\mathbf{1}_N \otimes I)\theta_U(t) + (\bar{U} \otimes I)e^{A_\varsigma t}(\tilde{U} \otimes I)\tilde{\xi}(0). \quad (4.105)$$

Define

$$\eta_i(t) = \begin{bmatrix} \eta_{ix}(t) \\ \eta_{iv}(t) \end{bmatrix} = (e_i^T\bar{U} \otimes I)e^{A_\varsigma t}(\tilde{U} \otimes I)\tilde{\xi}(0). \quad (4.106)$$

If Theorem 4.8 holds, then A_ς is asymptotically stable and $\lim_{t\to\infty}\eta_i(t) = 0$. Therefore, there exists a nonnegative constant α_ς such that

$$|\eta_{ix}(t)| \leqslant \alpha_\varsigma \left\|\tilde{\xi}(0)\right\|. \quad (4.107)$$

From (4.105) and (4.106), one knows $\tilde{\zeta}_i(t) = \theta_U(t) + \eta_i(t)$. Since $\tilde{\zeta}_i(t) = \zeta_i(t) - h_i(t)$, one has

$$x_i(t) - x_j(t) = h_{ix}(t) - h_{jx}(t) + \eta_{ix}(t) - \eta_{jx}(t). \tag{4.108}$$

From (4.108), it holds that

$$\left|x_i(t) - x_j(t)\right| \geqslant \left|h_{ix}(t) - h_{jx}(t)\right| - \left|\eta_{ix}(t) - \eta_{jx}(t)\right|. \tag{4.109}$$

Inequality (4.109) shows that if

$$\left|h_{ix}(t) - h_{jx}(t)\right| > \left|\eta_{ix}(t) - \eta_{jx}(t)\right|, \ \forall t \geqslant 0, \tag{4.110}$$

then the collision avoidance can be ensured. From (4.107) and (4.110), one sees that if for $\forall t \geqslant 0$, $\left|h_{ix}(t) - h_{jx}(t)\right| > 2\alpha_\varsigma \left\|\tilde{\zeta}(0)\right\|$, then (4.110) holds. Therefore, if the initial formation error is small enough and the formation satisfies (4.103), then collisions among UAVs can be avoided. Similar analysis can be done for the case that only (4.91) holds.

4.4.3 Quadrotor UAV Formation Platform

In this subsection, a quadrotor UAV formation platform which is used to demonstrate the theoretical results obtained in this section is introduced.

The quadrotor UAV formation platform consists of five quadrotor UAVs with flight control system (FCS) and one ground control station (GCS), which is displayed in Fig. 4.10. The quadrotor UAV frames together with accompanying brushless motors, propellers, and electrical speed controllers are provided by Xaircraft [35]. Each quadrotor UAV has a tip-to-tip wingspan of 65 cm, a weight of about 1600 g, a battery life of 12 min, and a maximum take-off weight of 1800 g.

Fig. 4.10 Quadrotor UAV formation platform

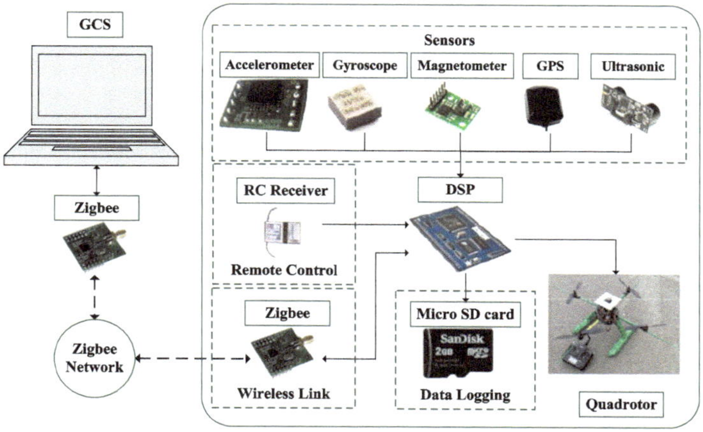

Fig. 4.11 Hardware structure of the quadrotor UAV system

To estimate the attitude and acceleration of the quadrotor, the FCS is equipped with three one-axis gyroscopes, a three-axis accelerometer, and a three-axis magnetometer. A global positioning system (GPS) module with an accuracy of 1.2 m circular error probable (CEP) is used to measure the position and velocity of each quadrotor at a rate of 10 Hz. When the quadrotor is near the ground, an ultrasonic range finder is adopted to get a more accurate height measurement. Moreover, a 2G micro SD card is mounted onboard for data logging. Zigbee modules are used for wireless communication among quadrotors and the GCS. An RC receiver is also mounted on the quadrotor in case of emergency. All the computations are performed on the onboard TMS320F28335 DSP running at 135 MHz. Through the Zigbee network, control commands can be sent to a specified quadrotor UAV or broadcast to all quadrotor UAVs, and the states of all quadrotor UAVs can be sent to the GCS. In addition, the states of all quadrotor UAVs can be monitored by the real-time display module on the GCS. Figure 4.11 illustrates the hardware structure of the quadrotor UAV system.

4.4.4 Simulation and Experimental Applications on Quadrotor UAVs

In this subsection, two applications on five quadrotor UAVs are shown to demonstrate the effectiveness of theoretical results in this section. Each quadrotor UAV interacts with its neighbors via Zigbee network, and the interaction topology of the quadrotor UAV swarm system in the applications is shown in Fig. 4.12. For simplicity, it is assumed that all quadrotor UAVs move in the XY plane; that is, $n = 2$.

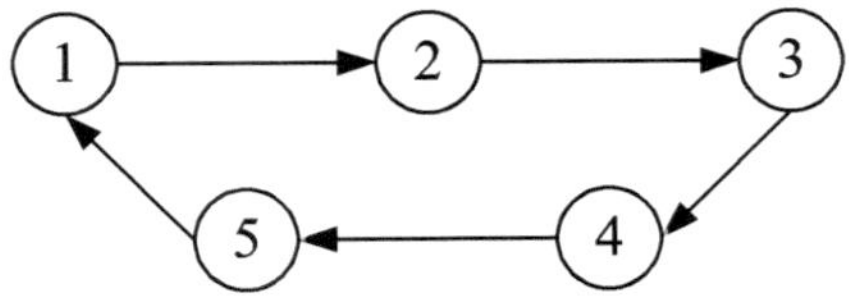

Fig. 4.12 Directed interaction topology G

The outer-loops for trajectory dynamics along X-axis and Y-axis are controlled by the formation controller (4.79), and the inner-loops for attitudes rotating around X-axis and Y-axis are controlled by PD controllers shown in [36]. The height and the yaw angle of each quadrotor UAV are specified to be constants. The attitude controller runs at 500 Hz, while the formation controller runs at 10 Hz. It should be pointed out that when quadrotors are achieving formation, they require neither remote control nor the GCS control.

Let $\zeta_i(t)=[x_{iX}(t), v_{iX}(t), x_{iY}(t), v_{iY}(t)]^T$ and $h_i(t)=[h_{ixX}(t), h_{ivX}(t), h_{ixY}(t), h_{ivY}(t)]^T$. Using Kronecker product, the dynamics of each quadrotor UAV in the case that $n = 2$ can be written as

$$\dot{\zeta}_i(t) = \left(I_2 \otimes B_1 B_2^T\right) \zeta_i(t) + (I_2 \otimes B_2)\, u_i(t).$$

Example 4.4 Consider the following time-varying formations

$$h_i(t) = \begin{bmatrix} r\left(\cos\left(\omega t + \frac{2\pi(i-1)}{5}\right) - 1\right) g_i(t) \\ -\omega r \sin\left(\omega t + \frac{2\pi(i-1)}{5}\right) g_i(t) \\ r \sin\left(\omega t + \frac{2\pi(i-1)}{5}\right) \\ \omega r \cos\left(\omega t + \frac{2\pi(i-1)}{5}\right) \end{bmatrix} \quad (i = 1, 2, \ldots, 5),$$

where $r = 7$ m, $\omega = 0.214$ rad/s, $g_i(t) = \text{sign}\left(\sin\left(\omega t/2 + \pi(i-1)/5\right)\right)$ $(i = 1, 2, \ldots, 5)$. From $h_i(t)$ $(i = 1, 2, \ldots, 5)$, one sees that if the quadrotor UAV swarm system achieves the desired time-varying formations, then the five quadrotor UAVs will follow a figure eight pattern while keeping a phase separation of $2\pi/5$ rad. It should be pointed out that it is different from the results in [33, 37, 38] that when the time-varying formation is achieved, the velocities of quadrotor UAVs are not identical. Due to the limitation of flight space and the requirement of performing flight experiments within a visual range, the eigenvalues of $B_2K_1 + B_1B_2^T$ are placed at $-0.6 + 1.28j$ and $-0.6 - 1.28j$ with $j^2 = -1$ by $K_1 = I_2 \otimes [-2, -1.2]$, which means that when the quadrotor UAV swarm system achieves time-varying formation, the formation reference is stationary. In order to avoid collisions, the initial states of quadrotor UAVs will be chosen to be near the trajectories of formations. It can be verified that condition (i) in Theorem 4.8 is satisfied. From Theorem 4.9, one can obtain a K_2 as follows:

$$K_2 = I_2 \otimes \begin{bmatrix} 0.3416 & 0.7330 \end{bmatrix}.$$

The initial states for the quardrotor UAVs are chosen as $\zeta_1(0) = [-0.16, 0.03, -0.07, -0.01]^T$, $\zeta_2(0) = [-4.92, -0.08, 6.38, -0.04]^T$, $\zeta_3(0) = [-12.37, -0.26, 4.08, -0.03]^T$, $\zeta_4(0) = [-12.73, 0.03, -4.56, -0.04]^T$, and $\zeta_5(0)=[-4.63, -0.05, -6.9, 0.02]^T$. Figures 4.13 and 4.14 show the state trajectories of the quadrotor UAV swarm system in simulation and experiment, and the predefined formation reference function $r(t)$ within 180 s, respectively, where the initial states of quadrotor UAVs and $r(0)$ are denoted by circles, and the final states of quadrotor UAVs and $r(t)$ are denoted by square, diamond, down triangle, up triangle, left triangle, and pentagram, respectively. Figure 4.15 depicts the snapshots of positions and velocities of the five quadrotor UAVs in the experiment, and $r(t)$ from 16 s to 21 s. Figure 4.16 shows a formation flight picture of five quadrotor UAVs in Example 4.4. From Figs. 4.13–4.16, one sees that the quadrotor UAV swarm system achieves the predefined time-varying formations in both simulation and experiment. The videos of the experiment in Example 4.4 can be found at https://youtu.be/wpRJMQvCQQQ or http://v.youku.com/v_show/id_XNjIwNjUyNDg4.html.

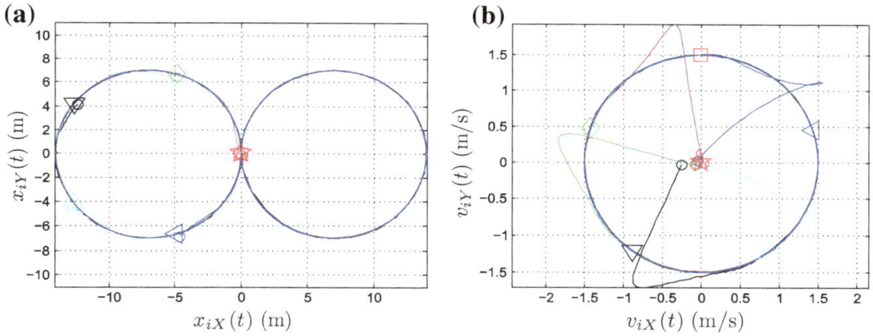

Fig. 4.13 Simulation state trajectories of five quadrotor UAVs and $r(t)$ in Example 4.4. **a** Position trajectories in simulation. **b** Velocity trajectories in simulation

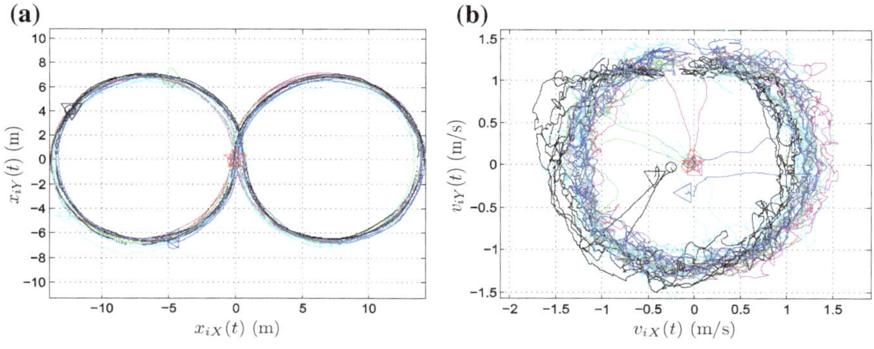

Fig. 4.14 Experimental state trajectories of five quadrotor UAVs and $r(t)$ in Example 4.4. **a** Position trajectories in experiment. **b** Velocity trajectories in experiment

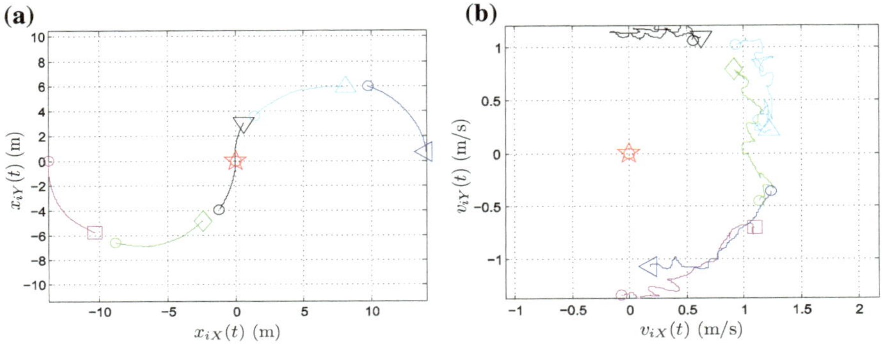

Fig. 4.15 Experimental snapshots of states and $r(t)$ for $t \in [4\text{s}, 49\text{s}]$ in Example 4.4. **a** Position snapshots in experiment. **b** Velocity snapshots in experiment

Fig. 4.16 A formation flight picture of five quadrotor UAVs in Example 4.4

Example 4.5 Consider the following time-varying formations

$$h_i(t) = \begin{bmatrix} r\sin(\omega t) + d\cos\left(\frac{2\pi(i-1)}{5}\right) \\ \omega r\cos(\omega t) \\ r\sin(2\omega t) + d\sin\left(\frac{2\pi(i-1)}{5}\right) \\ 2\omega r\cos(2\omega t) \end{bmatrix} \quad (i = 1, 2, \ldots, 5),$$

where $r = 7$ m, $d = 5$ m and $\omega = 0.143$ rad/s. From $h(t)$, one sees that if the quadrotor UAV swarm system achieves this time-varying formations, then the five quadrotor UAVs will keep a regular pentagon while flying along a Lissajous curve. Choose the same K_1 and K_2 as the ones in Example 4.4. It can be shown that the condition (i) in Theorem 4.8 is satisfied and K_2 can ensure that the UAV swarm system achieves the desired formation.

Choosing the initial states for five quadroror UAVs as $\zeta_1(0)=[5.19, 0, -0.04, 0]^T$, $\zeta_2(0) = [1.89, -0.12, 4.38, -0.11]^T$, $\zeta_3(0) = [-3.65, -0.29, 2.71, -0.05]^T$, $\zeta_4(0) = [-3.90, -0.18, -3.60, 0.08]^T$, $\zeta_5(0) = [2.00, -0.02, -5.13, 0.02]^T$. Figures 4.17 and 4.18 show the state trajectories of five quadrotor UAVs in simulation and experiment and the trajectory of $r(t)$ within 130 s. Figure 4.19b shows the snapshots of positions and velocities of the five quadrotor UAVs in the experiment, and $r(t)$ from 18 to 58 s. A formation flight picture of five quadrotor UAVs in Example 4.5 is depicted in Fig. 4.19. From Figs. 4.17, 4.18, 4.19 and 4.20, it can be seen that the quadrotor UAV swarm system achieved the predefined time-varying formations. The videos of the experiment in Example 4.5 can be found at https://youtu.be/wpRJMQvCQQQ or http://v.youku.com/v_show/id_XNjIwNjUyNDg4.html.

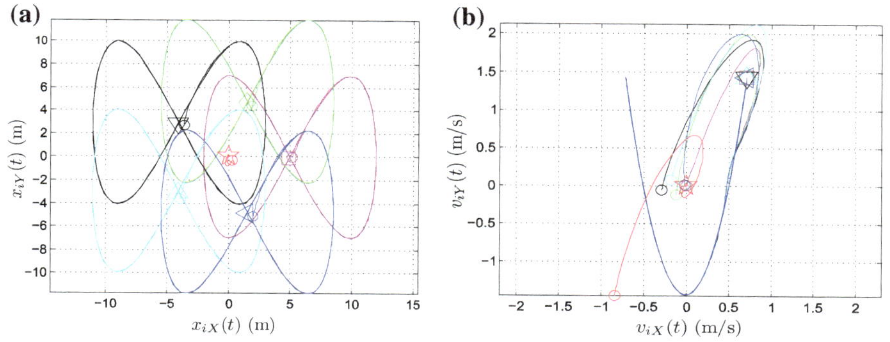

Fig. 4.17 Simulation state trajectories of five quadrotor UAVs and $r(t)$ in Example 4.5. **a** Position trajectories in simulation. **b** Velocity trajectories in simulation

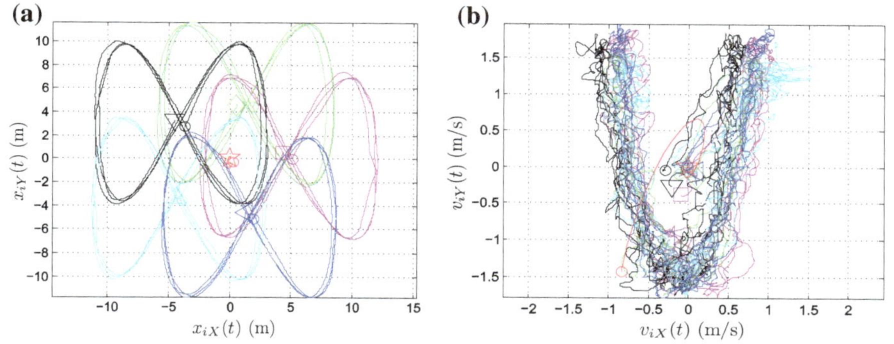

Fig. 4.18 Experimental state trajectories of five quadrotor UAVs and $r(t)$ in Example 4.5. **a** Position trajectories in experiment. **b** Velocity trajectories in experiment

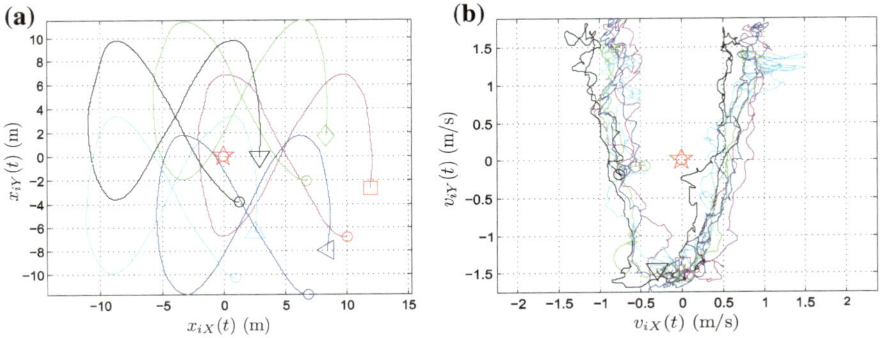

Fig. 4.19 Experimental snapshots of states and $r(t)$ for $t \in [4\text{s}, 49\text{s}]$ in Example 4.5. **a** Position snapshots in experiment. **b** Velocity snapshots in experiment

Fig. 4.20 A formation flight picture of five quadrotor UAVs in Example 4.5

4.5 Conclusions

This chapter first studied time-varying state formation control problems for high-order LTI swarm systems with time delays. Necessary and sufficient conditions for time-delayed swarm systems to achieve time-varying state formation, and for state formation feasibilities were proposed. An explicit expression for time-varying state formation reference function and approaches to assign the motion modes of the state formation reference were given. Approaches to expand the the feasible state formation set and to design the state formation protocol were proposed. The results on time-delayed swarm systems can be extended to swarm systems without time delays. Then based on static output feedback approaches, time-varying output formation problems for high-order LTI swarm systems were addressed. Necessary and sufficient conditions to achieve time-varying output formation, and necessary and sufficient conditions for output formation feasibility were presented. An explicit expression for time-varying output formation reference function, approaches to partially specify the motion modes of the output formation reference, and approaches to expand the feasible output formation set were proposed, respectively. Approaches to determine the gain matrices in the output formation protocol was also given. Finally, necessary and sufficient conditions for UAV swarm systems to achieve time-varying formations, and an explicit expression for the formation reference function were presented. Approaches to assign the motion modes of the formation reference and approaches to design the protocol were proposed. It was proven that by choosing appropriate initial states for each UAV, the collision among UAVs can be avoided. Based on a UAV formation platform consisting of five quadrotor UAVs, several time-varying formations experiments were performed in outdoor environment to demonstrate the theoretical results obtained in this chapter. The materials of this chapter are mainly based on [39–41].

It should be pointed out that state/output consensus problems can be regarded as special cases of state/output formation control problems. If $h(t) \equiv 0$, all the results in this chapter can be applied to deal with the corresponding consensus problems. Moreover, we have investigated time-varying state formation control problems for high-order LTI swarm systems with switching interaction topologies. Based on the approaches of state transformation, state space decomposition, and common Lyapunov function, necessary and sufficient conditions for swarm systems with switching interaction topologies to achieve time-varying state formation and for state formation feasibilities were proposed. Details on time-varying state formation control for high-order LTI swarm systems with switching interaction topologies can refer to the article [42] we have published and are omitted in this chapter.

References

1. Williamson WR, Abdel-Hafez MF, Rhee I et al (2007) An instrumentation system applied to formation flight. IEEE Trans Control Syst Technol 15(1):75–85
2. Nigam N, Bieniawski S, Kroo I et al (2012) Control of multiple UAVs for persistent surveillance: algorithm and flight test results. IEEE Trans Control Syst Technol 20(5):1236–1251
3. Kopfstedt T, Mukai M, Fujita M, et al (2008) Control of formations of UAVs for surveillance and reconnaissance missions. In: Proceedings of the 17th IFAC World Congress, pp 6–11
4. Han J, Xu Y, Di L et al (2013) Low-cost multi-UAV technologies for contour mapping of nuclear radiation field. J Intell Robot Syst 70(1–4):401–410
5. Pack DJ, DeLima P, Toussaint GJ et al (2009) Cooperative control of UAVs for localization of intermittently emitting mobile targets. IEEE Trans Syst Man Cybern B Cybern 39(4):959–970
6. Sivakumar A, Tan CKY (2010) UAV swarm coordination using cooperative control for establishing a wireless communications backbone. In: Proceedings of the 9th International Conference on Autonomous Agents and Multiagent Systems, pp 1157–1164
7. Wang PKC (1991) Navigation strategies for multiple autonomous mobile robots moving in formation. J Robotic Syst 8(2):177–195
8. Balch T, Arkin R (1998) Behavior-based formation control for multirobot teams. IEEE Trans Robot Autom 14(6):926–939
9. Lewis M, Tan K (1997) High precision formation control of mobile robots using virtual structures. Auton Robot 4(4):387–403
10. Beard RW, Lawton J, Hadaegh FY (2001) A coordination architecture for spacecraft formation control. IEEE Trans Control Syst Technol 9(6):777–790

11. Ren W (2007) Consensus strategies for cooperative control of vehicle formations. IET Control Theory Appl 1(2):505–512
12. Ren W, Sorensen N (2008) Distributed coordination architecture for multi-robot formation control. Robot Auton Syst 56(4):324–333
13. Xiao F, Wang L, Chen J et al (2009) Finite-time formation control for multi-agent systems. Automatica 45(11):2605–2611
14. Xie GM, Wang L (2009) Moving formation convergence of a group of mobile robots via decentralised information feedback. Int J Syst Sci 40(10):1019–1027
15. Liu CL, Tian YP (2009) Formation control of multi-agent systems with heterogeneous communication delays. Int J Syst Sci 40(6):627–636
16. Lafferriere G, Williams A, Caughman J et al (2005) Decentralized control of vehicle formations. Syst Control Lett 54(9):899–910
17. Fax JA, Murray RM (2004) Information flow and cooperative control of vehicle formations. IEEE Trans Autom Control 49(9):1465–1476
18. Porfiri M, Roberson DG, Stilwell DJ (2007) Tracking and formation control of multiple autonomous agents: a two-level consensus approach. Automatica 43(8):1318–1328
19. Lin ZY, Francis B, Maggiore M (2005) Necessary and sufficient graphical conditions for formation control of unicycles. IEEE Trans Autom Control 50(1):121–127
20. Ma CQ, Zhang JF (2012) On formability of linear continuous-time multi-agent systems. J Syst Sci Complex 25(1):13–29
21. Xi JX, Cai N, Zhong YS (2010) Consensus problems for high-order linear time-invariant swarm systems. Phys A 389(24):5619–5627
22. Zhang XM, Wu M, She JH et al (2005) Delay-dependent stabilization of linear systems with time-varying state and input delays. Automatica 41(8):1405–1412
23. Du B, Lam J, Shu Z et al (2009) A delay-partitioning projection approach to stability analysis of continuous systems with multiple delay components. IET Control Theory Appl 3(4):383–390
24. Xi JX, Shi ZY, Zhong YS (2011) Consensus analysis and design for high-order linear swarm systems with time-varying delays. Phys A 390(23–24):4114–4123
25. Ren W (2009) Collective motion from consensus with Cartesian coordinate coupling. IEEE Trans Autom Control 54(6):1330–1335
26. Lin P, Jia YM (2010) Distributed rotating formation control of multi-agent systems. Syst Control Lett 59(10):587–595
27. Xi JX, Shi ZY, Zhong YS (2012) Output consensus for high-order linear time-invariant swarm systems. Int J Control 85(4):350–360
28. He Y, Wang Q (2006) An improved ILMI method for static output feedback control with application to multivariable PID control. IEEE Trans Autom Control 51(10):1678–1683
29. Bayezit I, Fidan B (2013) Distributed cohesive motion control of flight vehicle formations. IEEE Trans Ind Electron 60(12):5763–5772
30. Karimoddini A, Lin H, Chen BM et al (2013) Hybrid three-dimensional formation control for unmanned helicopters. Automatica 49(2):424–433
31. Wang XH, Yadav V, Balakrishnan SN (2007) Cooperative UAV formation flying with obstacle/collision avoidance. IEEE Trans Control Syst Technol 15(4):672–679
32. Wang JN, Xin M (2013) Integrated optimal formation control of multiple unmanned aerial vehicles. IEEE Trans Control Syst Technol 21(5):1731–1744
33. Seo J, Kim Y, Kim S et al (2012) Consensus-based reconfigurable controller design for unmanned aerial vehicle formation flight. J Aerosp Eng 226(7):817–829
34. Zahreddine Z, El-Shehawey EF (1988) On the stability of a system of differential equations with complex coefficients. Indian J Pure Appl Math 19(10):963–972
35. XAircraft. http://www.xaircraft.com
36. Tayebi A, McGilvray S (2006) Attitude stabilization of a VTOL quadrotor aircraft. IEEE Trans Control Syst Technol 14(3):562–571
37. Abdessameud A, Tayebi A (2011) Formation control of VTOL unmanned aerial vehicles with communication delays. Automatica 47(11):2383–2394

38. Turpin M, Michael N, Kumar V (2012) Decentralized formation control with variable shapes for aerial robots. In: Proceedings of IEEE International Conference on Robotics and Automation, pp 23–30
39. Dong XW, Xi JX, Lu G et al (2014) Formation control for high-order linear time-invariant multi-agent systems with time delays. IEEE Trans Control Netw Syst 1(3):232–240
40. Dong XW, Shi ZY, Lu G et al (2015) Time-varying output formation control for high-order linear time-invariant swarm systems. Inf Sci 298(20):36–52
41. Dong XW, Yu BC, Shi ZY et al (2015) Time-varying formation control for unmanned aerial vehicles: theories and applications. IEEE Trans Control Syst Technol 23(1):340–348
42. Dong XW, Shi ZY, Lu G et al (2014) Time-varying formation control for high-order linear swarm systems with switching interaction topologies. IET Control Theory Appl 8(18):2162–2170

Chapter 5
Containment Control of Swarm Systems

Abstract This chapter studies output containment control problems for high-order linear time-invariant (LTI) swarm systems and state containment problems for high-order LTI singular swarm systems with time delays. First, a dynamic output containment protocol is presented. Necessary and sufficient conditions for swarm systems to achieve output containment are proposed. To ensure the scalability of the criteria, a sufficient condition which only includes two linear matrix inequality constraints independent of the number of agents is further presented. An approach independent of the number of agents is proposed to determine the gain matrices in the dynamic output containment protocols by solving an algebraic Riccati equation. Second, to eliminate impulse terms in singular swarm systems and ensure that the singular swarm systems can achieve containment, time-delayed protocols are presented for leaders and followers, respectively. By model transformation, containment problems of singular swarm systems are converted into stability problems of multiple low-dimensional time-delayed systems. In terms of linear matrix inequality, sufficient conditions are presented for time-delayed singular swarm systems to achieve state containment, which are independent of the number of agents. By using the method of changing variables, an approach is provided to determine the gain matrices in the protocols. Finally, numerical simulations are presented to demonstrate theoretical results.

5.1 Introduction

In Chaps. 3 and 4, it was assumed that there exist no leaders. In some practical applications, there exist multiple leaders which provide reference states or outputs for the whole swarm system. If it is required that the states/outputs of followers converge to the convex hull formed those of leaders, then containment control problems arise. Ji et al. [1] proposed a hybrid stop–go control strategy for first-order swarm systems to achieve containment. Meng et al. [2] presented nonlinear control protocols for rigid body swarm systems to achieve containment in a finite time. In [3, 4], containment problems for first-order and second-order swarm systems with switching

X. Dong, *Formation and Containment Control for High-order Linear Swarm Systems*, Springer Theses, DOI 10.1007/978-3-662-47836-3_5

interaction topologies were discussed. Cao et al. [5] applied containment theories to containment control of practical robot swarm systems. Liu et al. [6] proposed necessary and sufficient conditions for first-order and second-order swarm systems to achieve containment. Containment problems for second-order swarm systems with random switching interaction topologies were studied in [7]. Containment control problems for high-order LTI swarm systems were investigated in [8, 9].

For containment control of swarm systems, the following problems need to be dealt with. First, in most of the previous results, the dynamics of each agent is first order or second order. Containment control problems for high-order swarm systems need further research, and output containment control problems have not been studied before. Second, in some practical applications, the dynamics of each agent can only be described by singular model, and time delays are inevitable. Containment control problems for high-order LTI singular swarm systems with time delays are still open.

Output containment control problems for high-order swarm systems and state containment control problems for high-order singular swarm systems with time delays are addressed in this chapter. The main results are summarized as follows. First, an output containment control protocol is constructed based on dynamic output feedback approaches. Necessary and sufficient conditions for swarm systems to achieve output containment are proposed and approaches to design the output containment protocol are presented. Second, a time-delayed state containment control protocol is presented for singular swarm systems. Sufficient conditions for high-order singular swarm systems with time delays to achieve state containment are proposed. Using the method of changing variables, an approach is given to determine the gain matrices in the protocols. The results on containment control of singular swarm systems with time delays can be applied to solve the consensus tracking problems for singular swarm systems with time delays, containment control problems, and consensus tracking problems for high-order swarm systems with time delays.

The rest of this chapter is organized as follows. In Sect. 5.2, output containment control problems for high-order swarm systems are studied. Section 5.3 investigates state containment control problems for high-order singular swarm systems with time delays. Section 5.4 concludes the whole work of this chapter.

5.2 Dynamic Output Containment Control

In this section, output containment control problems for high-order swarm systems are investigated using dynamic output feedback control approaches. Necessary and sufficient conditions to achieve output containment and approaches to design the output containment protocol are derived.

5.2.1 Problem Description

Consider a high-order LTI swarm system with N agents. The dynamics of each agent is described by

$$\begin{cases} \dot{x}_i(t) = Ax_i(t) + Bu_i(t), \\ y_i(t) = Cx_i(t), \end{cases} \tag{5.1}$$

where $i \in \{1, 2, \ldots, N\}$, $x_i(t) \in \mathbb{R}^n$ is the state, $y_i(t) \in \mathbb{R}^q$ is the output, and $u_i(t) \in \mathbb{R}^m$ is the control input. It is assumed that C is of full row rank.

Definition 5.1 An agent is called a *leader* if it has no neighbors. An agent is called a *follower* if it has at least one neighbor.

Assume that in swarm system (5.1) there are M $(M < N)$ followers with states $x_k(t)$ $(k = 1, 2, \ldots, M)$ and $N - M$ leaders with states $x_i(t)$ $(i = M + 1, M + 2, \ldots, N)$. Let $\mathscr{F} = \{1, 2, \ldots, M\}$ and $\mathscr{L} = \{M + 1, M + 2, \ldots, N\}$ be the follower subscript set and leader subscript set, respectively. Under Definition 5.1, the Laplacian matrix corresponding to G has the following form:

$$L = \begin{bmatrix} L_1 & L_2 \\ 0 & 0 \end{bmatrix},$$

where $L_1 \in \mathbb{R}^{M\times M}$ and $L_2 \in \mathbb{R}^{M\times(N-M)}$.

Assumption 1 For each follower, there exists at least one leader that has a directed path to it.

Under Assumption 1, the following lemma holds.

Lemma 5.1 ([2]) *If the interaction topology satisfies Assumption 1, then all the eigenvalues of L_1 have positive real parts, each entry of $-L_1^{-1}L_2$ is nonnegative, and each row of $-L_1^{-1}L_2$ has a sum equal to one.*

Definition 5.2 Swarm system (5.1) is said to *achieve output containment* if for any given bounded initial states and any $k \in \mathscr{F}$, there exist nonnegative constants $\beta_{k,j}$ $(j \in \mathscr{L})$ satisfying $\sum_{j=M+1}^{N} \beta_{k,j} = 1$ such that

$$\lim_{t\to\infty} \left(y_k(t) - \sum_{j=M+1}^{N} \beta_{k,j} y_j(t) \right) = 0. \tag{5.2}$$

Consider the following dynamic output containment protocols for followers and leaders, respectively,

$$\begin{cases} \dot{z}_i(t) = K_1 z_i(t) + K_2 \sum_{j\in N_i} w_{ij} \left(z_i(t) - z_j(t)\right) + K_3 \sum_{j\in N_i} w_{ij} \left(y_i(t) - y_j(t)\right), i \in \mathscr{F}, \\ u_i(t) = K_4 z_i(t), i \in \mathscr{F}, \end{cases} \tag{5.3}$$

$$\begin{cases} \dot{z}_i(t) = K_1 z_i(t), i \in \mathcal{L}, \\ u_i(t) = K_4 z_i(t), i \in \mathcal{L}, \end{cases} \tag{5.4}$$

where $z_i(t) \in \mathbb{R}^{\bar{n}}$ $(i = 1, 2, \ldots, N)$ are the states of the protocols, and K_1, K_2, K_3, and K_4 are constant gain matrices with appropriate dimensions.

Let

$$z_{\mathcal{F}}(t) = [z_1^T(t), z_2^T(t), \ldots, z_M^T(t)]^T, \ x_{\mathcal{F}}(t) = [x_1^T(t), x_2^T(t), \ldots, x_M^T(t)]^T,$$

$$y_{\mathcal{F}}(t) = [y_1^T(t), y_2^T(t), \ldots, y_M^T(t)]^T, \ z_{\mathcal{L}}(t) = [z_{M+1}^T(t), z_{M+2}^T(t), \ldots, z_N^T(t)]^T,$$

$$x_{\mathcal{L}}(t) = [x_{M+1}^T(t), x_{M+2}^T(t), \ldots, x_N^T(t)]^T, \ y_{\mathcal{L}}(t) = [y_{M+1}^T(t), y_{M+2}^T(t), \ldots, y_N^T(t)]^T.$$

Under protocols (5.3) and (5.4), the dynamics of swarm system (5.1) can be written in a compact form as

$$\begin{cases} \dot{z}_{\mathcal{F}}(t) = (I_M \otimes K_1 + L_1 \otimes K_2) z_{\mathcal{F}}(t) + (L_2 \otimes K_2) z_{\mathcal{L}}(t) \\ \qquad\qquad +(L_1 \otimes K_3) y_{\mathcal{F}}(t) + (L_2 \otimes K_3) y_{\mathcal{L}}(t), \\ \dot{x}_{\mathcal{F}}(t) = (I_M \otimes A) x_{\mathcal{F}}(t) + (I_M \otimes BK_4) z_{\mathcal{F}}(t), \\ y_{\mathcal{F}}(t) = (I_M \otimes C) x_{\mathcal{F}}(t), \end{cases} \tag{5.5}$$

$$\begin{cases} \dot{z}_{\mathcal{L}}(t) = (I_{N-M} \otimes K_1) z_{\mathcal{L}}(t), \\ \dot{x}_{\mathcal{L}}(t) = (I_{N-M} \otimes A) x_{\mathcal{L}}(t) + (I_{N-M} \otimes BK_4) z_{\mathcal{L}}(t), \\ y_{\mathcal{L}}(t) = (I_{N-M} \otimes C) x_{\mathcal{L}}(t). \end{cases} \tag{5.6}$$

In this section, the following two problems for swarm system (5.1) under dynamic protocols (5.3) and (5.4) are mainly addressed: (i) under what conditions output containment can be achieved; and (ii) how to determine the gain matrices in dynamic protocols (5.3) and (5.4) to achieve output containment.

5.2.2 Dynamic Output Containment Analysis

Apply the same $\bar{C}$ and T defined in Section 4.3.2 to the subsystems with states $x_{\mathcal{F}}(t)$ and $x_{\mathcal{L}}(t)$, respectively. Let $\bar{y}_i(t) = \bar{C} x_i(t)$ $(i = 1, 2, \ldots, N)$, $\bar{y}_{\mathcal{F}}(t) = [\bar{y}_1^T(t), \bar{y}_2^T(t), \ldots, \bar{y}_M^T(t)]^T$ and $\bar{y}_{\mathcal{L}}(t) = [\bar{y}_{M+1}^T(t), \bar{y}_{M+2}^T(t), \ldots, \bar{y}_N^T(t)]^T$. Then swarm system (5.5) and (5.6) can be decomposed into

$$\begin{cases} \dot{z}_{\mathcal{F}}(t) = (I_M \otimes K_1 + L_1 \otimes K_2)\, z_{\mathcal{F}}(t) + (L_2 \otimes K_2)\, z_{\mathcal{L}}(t) + (L_1 \otimes K_3)\, y_{\mathcal{F}}(t) \\ \qquad\qquad + (L_2 \otimes K_3)\, y_{\mathcal{L}}(t), \\ \dot{y}_{\mathcal{F}}(t) = \left(I_M \otimes \bar{A}_{11}\right) y_{\mathcal{F}}(t) + \left(I_M \otimes \bar{A}_{12}\right) \bar{y}_{\mathcal{F}}(t) + \left(I_M \otimes \bar{B}_1 K_4\right) z_{\mathcal{F}}(t), \\ \dot{\bar{y}}_{\mathcal{F}}(t) = \left(I_M \otimes \bar{A}_{21}\right) y_{\mathcal{F}}(t) + \left(I_M \otimes \bar{A}_{22}\right) \bar{y}_{\mathcal{F}}(t) + \left(I_M \otimes \bar{B}_2 K_4\right) z_{\mathcal{F}}(t), \end{cases} \tag{5.7}$$

$$\begin{cases} \dot{z}_{\mathscr{L}}(t) = (I_{N-M} \otimes K_1)\, z_{\mathscr{L}}(t), \\ \dot{y}_{\mathscr{L}}(t) = \left(I_{N-M} \otimes \bar{A}_{11}\right) y_{\mathscr{L}}(t) + \left(I_{N-M} \otimes \bar{A}_{12}\right) \bar{y}_{\mathscr{L}}(t) + \left(I_{N-M} \otimes \bar{B}_1 K_4\right) z_{\mathscr{L}}(t), \\ \dot{\bar{y}}_{\mathscr{L}}(t) = \left(I_{N-M} \otimes \bar{A}_{21}\right) y_{\mathscr{L}}(t) + \left(I_{N-M} \otimes \bar{A}_{22}\right) \bar{y}_{\mathscr{L}}(t) + \left(I_{N-M} \otimes \bar{B}_2 K_4\right) z_{\mathscr{L}}(t). \end{cases} \quad (5.8)$$

In (5.7) and (5.8), the subsystems with states $y_{\mathscr{F}}(t)$ and $y_{\mathscr{L}}(t)$ can be used to analyze the output containment problems. It can be found that only the observable components of $(\bar{A}_{22}, \bar{A}_{12})$ influence the subsystems with states $y_F(t)$ and $y_J(t)$. Therefore, the observability decomposition of $(\bar{A}_{22}, \bar{A}_{12})$ is presented first using the matrix $\tilde{T}$ defined in Sect 4.3.3. Let $\hat{y}_i(t) = \tilde{T}^{-1}\bar{y}_i(t) = [\hat{y}_{io}^T(t), \hat{y}_{i\bar{o}}^T(t)]^T$ ($i = 1, 2, \ldots, N$),

$$\hat{y}_{\mathscr{F}o}(t) = [\hat{y}_{1o}^T(t), \hat{y}_{2o}^T(t), \ldots, \hat{y}_{Mo}^T(t)]^T,\ \hat{y}_{\mathscr{F}\bar{o}}(t) = [\hat{y}_{1\bar{o}}^T(t), \hat{y}_{2\bar{o}}^T(t), \ldots, \hat{y}_{M\bar{o}}^T(t)]^T,$$

$$\hat{y}_{\mathscr{L}o}(t) = [\hat{y}_{(M+1)o}^T(t), \hat{y}_{(M+2)o}^T(t), \ldots, \hat{y}_{No}^T(t)]^T, \hat{y}_{\mathscr{L}\bar{o}}(t) = [\hat{y}_{(M+1)\bar{o}}^T(t), \hat{y}_{(M+2)\bar{o}}^T(t), \ldots, \hat{y}_{N\bar{o}}^T(t)]^T.$$

Then systems (5.7) and (5.8) can be further converted into

$$\begin{cases} \dot{z}_{\mathscr{F}}(t) = (I_M \otimes K_1 + L_1 \otimes K_2)\, z_{\mathscr{F}}(t) + (L_2 \otimes K_2)\, z_{\mathscr{L}}(t) + (L_1 \otimes K_3)\, y_{\mathscr{F}}(t) \\ \qquad + (L_2 \otimes K_3)\, y_{\mathscr{L}}(t), \\ \dot{y}_{\mathscr{F}}(t) = \left(I_M \otimes \bar{A}_{11}\right) y_{\mathscr{F}}(t) + \left(I_M \otimes \tilde{E}_1\right) \hat{y}_{\mathscr{F}o}(t) + \left(I_M \otimes \bar{B}_1 K_4\right) z_{\mathscr{F}}(t), \\ \dot{\hat{y}}_{\mathscr{F}o}(t) = \left(I_M \otimes \tilde{F}_1\right) y_{\mathscr{F}}(t) + \left(I_M \otimes \tilde{D}_1\right) \hat{y}_{\mathscr{F}o}(t) + \left(I_M \otimes \tilde{B}_1 K_4\right) z_{\mathscr{F}}(t), \\ \dot{\hat{y}}_{\mathscr{F}\bar{o}}(t) = \left(I_M \otimes \tilde{F}_2\right) y_{\mathscr{F}}(t) + \left(I_M \otimes \tilde{D}_2\right) \hat{y}_{\mathscr{F}o}(t) + \left(I_M \otimes \tilde{D}_3\right) \hat{y}_{\mathscr{F}\bar{o}}(t) \\ \qquad + \left(I_M \otimes \tilde{B}_2 K_4\right) z_{\mathscr{F}}(t), \end{cases} \quad (5.9)$$

$$\begin{cases} \dot{z}_{\mathscr{L}}(t) = (I_{N-M} \otimes K_1)\, z_{\mathscr{L}}(t), \\ \dot{y}_{\mathscr{L}}(t) = \left(I_{N-M} \otimes \bar{A}_{11}\right) y_{\mathscr{L}}(t) + \left(I_{N-M} \otimes \tilde{E}_1\right) \hat{y}_{\mathscr{L}o}(t) + \left(I_{N-M} \otimes \bar{B}_1 K_4\right) z_{\mathscr{L}}(t), \\ \dot{\hat{y}}_{\mathscr{L}o}(t) = \left(I_{N-M} \otimes \tilde{F}_1\right) y_{\mathscr{L}}(t) + \left(I_{N-M} \otimes \tilde{D}_1\right) \hat{y}_{\mathscr{L}o}(t) + \left(I_{N-M} \otimes \tilde{B}_1 K_4\right) z_{\mathscr{L}}(t), \\ \dot{\hat{y}}_{\mathscr{L}\bar{o}}(t) = \left(I_{N-M} \otimes \tilde{F}_2\right) y_{\mathscr{L}}(t) + \left(I_{N-M} \otimes \tilde{D}_2\right) \hat{y}_{\mathscr{L}o}(t) + \left(I_{N-M} \otimes \tilde{D}_3\right) \hat{y}_{\mathscr{L}\bar{o}}(t) \\ \qquad + \left(I_{N-M} \otimes \tilde{B}_2 K_4\right) z_{\mathscr{L}}(t). \end{cases} \quad (5.10)$$

Let $U_{\mathscr{F}} \in \mathbb{C}^{M \times M}$ be a nonsingular matrix such that $U_{\mathscr{F}}^{-1} L_1 U_{\mathscr{F}} = \bar{\Lambda}_{\mathscr{F}}$, where $\bar{\Lambda}_{\mathscr{F}}$ is an upper-triangular matrix with λ_i ($i = 1, 2, \ldots, M$) as its diagonal entries and $\mathrm{Re}(\lambda_1) \leq \mathrm{Re}(\lambda_2) \leq \cdots \leq \mathrm{Re}(\lambda_M)$. The following theorem presents necessary and sufficient conditions for swarm system (5.1) under dynamic protocols (5.3) and (5.4) to achieve output containment.

Theorem 5.1 *For any given bounded initial states, swarm system (5.1) under dynamic protocols (5.3) and (5.4) achieves output containment if and only if for all* $i \in \{1, 2, \ldots, M\}$ *the observable modes of* $\left(\Upsilon_{\mathscr{F}i}, \begin{bmatrix} 0 & I & 0 \end{bmatrix}\right)$ *are asymptotically stable, where*

$$\Upsilon_{\mathcal{F}i} = \begin{bmatrix} K_1 + \lambda_i K_2 & \lambda_i K_3 & 0 \\ \bar{B}_1 K_4 & \bar{A}_{11} & \tilde{E}_1 \\ \tilde{B}_1 K_4 & \tilde{F}_1 & \tilde{D}_1 \end{bmatrix}.$$

Proof Sufficiency: Let $\phi_{\mathcal{F}}(t) = [z_{\mathcal{F}}^T(t), y_{\mathcal{F}}^T(t), \hat{y}_{\mathcal{F}o}^T(t)]^T$, $\phi_{\mathcal{L}}(t) = [z_{\mathcal{L}}^T(t), y_{\mathcal{L}}^T(t), \hat{y}_{\mathcal{L}o}^T(t)]^T$ and $\phi(t) = [\phi_{\mathcal{F}}^T(t), \phi_{\mathcal{L}}^T(t)]^T$. From (5.9) and (5.10), one has

$$\dot{\phi}(t) = \begin{bmatrix} \Phi_1 & \Phi_2 \\ 0 & \Phi_3 \end{bmatrix} \phi(t), \tag{5.11}$$

where

$$\Phi_1 = \begin{bmatrix} I_M \otimes K_1 + L_1 \otimes K_2 & L_1 \otimes K_3 & 0 \\ I_M \otimes \bar{B}_1 K_4 & I_M \otimes \bar{A}_{11} & I_M \otimes \tilde{E}_1 \\ I_M \otimes \tilde{B}_1 K_4 & I_M \otimes \tilde{F}_1 & I_M \otimes \tilde{D}_1 \end{bmatrix}, \quad \Phi_2 = \begin{bmatrix} L_2 \otimes K_2 & L_2 \otimes K_3 & 0 \\ 0 & 0 & 0 \\ 0 & 0 & 0 \end{bmatrix},$$

$$\Phi_3 = \begin{bmatrix} I_{N-M} \otimes K_1 & 0 & 0 \\ I_{N-M} \otimes \bar{B}_1 K_4 & I_{N-M} \otimes \bar{A}_{11} & I_{N-M} \otimes \tilde{E}_1 \\ I_{N-M} \otimes \tilde{B}_1 K_4 & I_{N-M} \otimes \tilde{F}_1 & I_{N-M} \otimes \tilde{D}_1 \end{bmatrix}.$$

The solution to Eq. (5.11) can be written as

$$\phi(t) = \begin{bmatrix} e^{\Phi_1 t} & \mathcal{L}^{-1}\left((sI - \Phi_1)^{-1}\Phi_2(sI - \Phi_3)^{-1}\right) \\ 0 & e^{\Phi_3 t} \end{bmatrix} \phi(0), \tag{5.12}$$

where $\mathcal{L}^{-1}$ is the inverse Laplace transform and s is the Laplace operator.

It can be shown that

$$(sI - \Phi_1)^{-1}\Phi_2(sI - \Phi_3)^{-1} = (sI - \Phi_1)^{-1}\Theta_L - \Theta_L(sI - \Phi_3)^{-1}.$$

where

$$\Theta_L = \begin{bmatrix} L_1^{-1}L_2 \otimes I & 0 & 0 \\ 0 & L_1^{-1}L_2 \otimes I & 0 \\ 0 & 0 & L_1^{-1}L_2 \otimes I \end{bmatrix}.$$

From (5.12), one has

$$y_{\mathcal{F}}(t) = [0\ I\ 0]\, e^{\Phi_1 t}\left(\phi_{\mathcal{F}}(0) + \left(L_1^{-1}L_2 \otimes I\right)\phi_{\mathcal{L}}(0)\right) + \left(-L_1^{-1}L_2 \otimes I\right) y_{\mathcal{L}}(t), \tag{5.13}$$

$$y_{\mathcal{L}}(t) = [0\ I\ 0]\, e^{\Phi_3 t}\phi_{\mathcal{L}}(0). \tag{5.14}$$

If the observable modes of $(\Upsilon_{\mathscr{F}i}, [\,0\ I\ 0\,])$ $(i = 1, 2, \ldots, M)$ are asymptotically stable, by the structure of $U_{\mathscr{F}}$ and $\bar{\Lambda}_{\mathscr{F}}$, it can be obtained that $\lim_{t\to\infty}\left[\,0\ I\ 0\,\right]e^{\Phi_1 t} = 0$. Therefore,

$$\lim_{t\to\infty}\left(y_{\mathscr{F}}(t) - \left(-L_1^{-1}L_2 \otimes I\right) y_{\mathscr{L}}(t)\right) = 0. \tag{5.15}$$

From (5.15) and Lemma 5.1, one knows that swarm system (5.1) under dynamic protocols (5.3) and (5.4) achieves output containment.

Necessity: The necessity is proven by contradiction. Choose $\phi_{\mathscr{L}}(0) = 0$ or $\phi_{\mathscr{L}}(0)$ in the unobservable subspace of $(\Phi_3, [\,0\ I\ 0\,])$. In this case, $y_{\mathscr{L}}(t) \equiv 0$. If swarm system (5.1) under dynamic protocols (5.3) and (5.4) achieves output containment but subsystem $(\Upsilon_{\mathscr{F}i}, [\,0\ I\ 0\,])$ has at least one observable mode which is not asymptotically stable for some $i \in \{1, 2, \ldots, M\}$, then subsystem $(\Phi_1, [\,0\ I\ 0\,])$ has at least one observable mode that is not asymptotically stable. Hence, one can find nonzero $\phi_{\mathscr{F}}(0)$ such that $\lim_{t\to\infty} y_{\mathscr{F}}(t) \neq 0$; that is, swarm system (5.1) under dynamic protocols (5.3) and (5.4) does not achieve output containment. This results a contradiction. Therefore, the condition that the observable modes of $(\Upsilon_{\mathscr{F}i}, [\,0\ I\ 0\,])$ are asymptotically stable is required. The proof for Theorem 5.1 is completed.

Remark 5.1 From the structure of $\Upsilon_{\mathscr{F}i}$ $(i = 1, 2, \ldots, M)$, one sees that the output containment of swarm systems depends on the dynamics of each agent, the observable components of $(\bar{A}_{22}, \bar{A}_{12})$, the interaction topology, and K_i $(i = 1, 2, 3, 4)$.

It is well known that one typical feature of swarm systems is of large scale. For swarm systems with a very large M, it may be time costly to check the conditions in Theorem 5.1. By encapsulating λ_i $(i = 1, 2, \ldots, M)$ into a convex set, a sufficient condition with less calculation complexity is obtained in the following theorem.

Let

$$\bar{\psi}_1 = \begin{bmatrix} K_1 & 0 & 0 \\ \bar{B}_1 K_4 & \bar{A}_{11} & \tilde{E}_1 \\ \tilde{B}_1 K_4 & \tilde{F}_1 & \tilde{D}_1 \end{bmatrix}, \quad \bar{\psi}_2 = \begin{bmatrix} K_2 & K_3 & 0 \\ 0 & 0 & 0 \\ 0 & 0 & 0 \end{bmatrix}.$$

Then $\Upsilon_{\mathscr{F}i} = \bar{\psi}_1 + \lambda_i \bar{\psi}_2$. Define $\bar{\lambda}_{1,2} = \mathrm{Re}(\lambda_1) \pm j\mu_{\mathscr{F}}$ and $\bar{\lambda}_{3,4} = \mathrm{Re}(\lambda_M) \pm j\mu_{\mathscr{F}}$, where $\mu_{\mathscr{F}} = \max\{\mathrm{Im}(\lambda_i), i \in \mathscr{F}\}$.

Theorem 5.2 *For any given bounded initial states, swarm system (5.1) under dynamic protocols (5.3) and (5.4) achieves output containment if there exists a matrix* $R_{\mathscr{F}} = R_{\mathscr{F}}^T > 0$ *such that*

$$\left(\Lambda_{\bar{\psi}_1} + \Psi_{\bar{\lambda}_i}\Lambda_{\bar{\psi}_2}\right)^T \Lambda_{R_{\mathscr{F}}} + \Lambda_{R_{\mathscr{F}}}^T \left(\Lambda_{\bar{\psi}_1} + \Psi_{\bar{\lambda}_i}\Lambda_{\bar{\psi}_2}\right) < 0\ (i = 1, 3). \tag{5.16}$$

Proof Let $\Pi_{\mathscr{F}i} = \left(\Lambda_{\bar{\psi}_1} + \Psi_{\lambda_i}\Lambda_{\bar{\psi}_2}\right)^T \Lambda_{R_{\mathscr{F}}} + \Lambda_{R_{\mathscr{F}}}^T \left(\Lambda_{\bar{\psi}_1} + \Psi_{\lambda_i}\Lambda_{\bar{\psi}_2}\right)$ $(i \in \mathscr{F})$. Then one can obtain

$$\Pi_{\mathscr{F}i} = \Omega_{\mathscr{F}0} + \mathrm{Re}(\lambda_i)\Omega_{\mathscr{F}1} + \mathrm{Im}(\lambda_i)\Omega_{\mathscr{F}2},$$

where

$$\Omega_{\mathscr{F}0} = \begin{bmatrix} \bar{\psi}_1^T R_{\mathscr{F}} + R_{\mathscr{F}}^T \bar{\psi}_1 & 0 \\ 0 & \bar{\psi}_1^T R_{\mathscr{F}} + R_{\mathscr{F}}^T \bar{\psi}_1 \end{bmatrix},$$

$$\Omega_{\mathscr{F}1} = \begin{bmatrix} \bar{\psi}_2^T R_{\mathscr{F}} + R_{\mathscr{F}}^T \bar{\psi}_2 & 0 \\ 0 & \bar{\psi}_2^T R_{\mathscr{F}} + R_{\mathscr{F}}^T \bar{\psi}_2 \end{bmatrix},$$

$$\Omega_{\mathscr{F}2} = \begin{bmatrix} 0 & \bar{\psi}_2^T R_{\mathscr{F}} - R_{\mathscr{F}}^T \bar{\psi}_2 \\ -\bar{\psi}_2^T R_{\mathscr{F}} + R_{\mathscr{F}}^T \bar{\psi}_2 & 0 \end{bmatrix}.$$

Similarly, let $\tilde{\Pi}_{\mathscr{F}i} = \left(\Lambda_{\bar{\psi}_1} + \Psi_{\bar{\lambda}_i}\Lambda_{\bar{\psi}_2}\right)^T \Lambda_{R_{\mathscr{F}}} + \Lambda_{R_{\mathscr{F}}}^T \left(\Lambda_{\bar{\psi}_1} + \Psi_{\bar{\lambda}_i}\Lambda_{\bar{\psi}_2}\right)$ $(i = 1, 2, 3, 4)$. Then it follows

$$\tilde{\Pi}_{\mathscr{F}i} = \Omega_{\mathscr{F}0} + \mathrm{Re}(\bar{\lambda}_i)\Omega_{\mathscr{F}1} + \mathrm{Im}(\bar{\lambda}_i)\Omega_{\mathscr{F}2}.$$

Let

$$T_{\mathscr{F}} = \begin{bmatrix} 0 & I \\ I & 0 \end{bmatrix}.$$

Since $\mathrm{Im}\left(\bar{\lambda}_1\right) = -\mathrm{Im}\left(\bar{\lambda}_2\right)$ and $\mathrm{Im}\left(\bar{\lambda}_3\right) = -\mathrm{Im}\left(\bar{\lambda}_4\right)$, one has $\tilde{\Pi}_{\mathscr{F}1} = T_{\mathscr{F}}\tilde{\Pi}_{\mathscr{F}2}T_{\mathscr{F}}^{-1}$ and $\tilde{\Pi}_{\mathscr{F}3} = T_{\mathscr{F}}\tilde{\Pi}_{\mathscr{F}4}T_{\mathscr{F}}^{-1}$. Therefore, if $\tilde{\Pi}_{\mathscr{F}1} < 0$ and $\tilde{\Pi}_{\mathscr{F}3} < 0$, then $\tilde{\Pi}_{\mathscr{F}2} < 0$ and $\tilde{\Pi}_{\mathscr{F}4} < 0$. By Lemma 4.3, one knows that if condition (5.16) holds, then $\Pi_{\mathscr{F}i} < 0$ $(i \in \mathscr{F})$.

Consider the stability of the following subsystem:

$$\dot{\varphi}_{\mathscr{F}i}(t) = \left(\Lambda_{\bar{\psi}_1} + \Psi_{\lambda_i}\Lambda_{\bar{\psi}_2}\right)\varphi_{\mathscr{F}i}(t) \ (i \in \mathscr{F}). \tag{5.17}$$

Choose the Lyapunov functional candidate as follows:

$$V_i(t) = \varphi_{\mathscr{F}i}^H(t)\Lambda_{R_{\mathscr{F}}}\varphi_{\mathscr{F}i}(t) \ (i \in \mathscr{F}).$$

Taking the time derivative of $V_i(t)$ along the trajectory of (5.17), one has

$$\dot{V}_i(t) = \varphi_{\mathscr{F}i}^H(t)\left(\left(\Lambda_{\bar{\psi}_1} + \Psi_{\lambda_i}\Lambda_{\bar{\psi}_2}\right)^T \Lambda_{R_{\mathscr{F}}} + \Lambda_{R_{\mathscr{F}}}\left(\Lambda_{\bar{\psi}_1} + \Psi_{\lambda_i}\Lambda_{\bar{\psi}_2}\right)\right)\varphi_{\mathscr{F}i}(t).$$

Since $\Pi_{\mathscr{F}i} < 0$ $(i \in \mathscr{F})$, one gets $\dot{V}_i(t) < 0$; that is, $\Lambda_{\bar{\psi}_1} + \Psi_{\lambda_i}\Lambda_{\bar{\psi}_2}$ $(i \in \mathscr{F})$ is Hurwitz. By the decomposition of real and imaginary parts, it can be verified that if $\Lambda_{\bar{\psi}_1} + \Psi_{\lambda_i}\Lambda_{\bar{\psi}_2}$ $(i \in \mathscr{F})$ is Hurwitz, then $\Upsilon_{\mathscr{F}i}$ $(i \in \mathscr{F})$ is Hurwitz. Therefore, the observable modes of $(\Upsilon_{\mathscr{F}i}, [\,0\ I\ 0\,])$ are asymptotically stable. From Theorem 5.1, one knows that swarm system (5.1) under dynamic protocols (5.3) and (5.4) achieves output containment. The proof of Theorem 5.2 is completed.

Remark 5.2 The condition (5.16) in Theorem 5.2 only includes two LMI constraints independent of the number of agents. As a result, for swarm systems with a huge number of followers, the calculation efficiency can be improved significantly, although the conclusion of Theorem 5.2 may be conservative compared with that of Theorem 5.1.

5.2.3 Dynamic Output Containment Protocol Design

In this subsection, an approach to determine K_i $(i = 1, 2, 3, 4)$ in dynamic protocols (5.3) and (5.4) for swarm system (5.1) to achieve output containment is proposed.

For simplicity of expression, let

$$A_{\mathscr{F}} = \begin{bmatrix} \bar{A}_{11} & \tilde{E}_1 \\ \tilde{F}_1 & \tilde{D}_1 \end{bmatrix},\ B_{\mathscr{F}} = \begin{bmatrix} \bar{B}_1 \\ \tilde{B}_1 \end{bmatrix},\ C_{\mathscr{F}} = \begin{bmatrix} I & 0 \end{bmatrix},$$

then $\Upsilon_{\mathscr{F}i}$ $(i \in \mathscr{F})$ can be rewritten as

$$\Upsilon_{\mathscr{F}i} = \begin{bmatrix} K_1 + \lambda_i K_2 & \lambda_i K_3 C_{\mathscr{F}} \\ B_{\mathscr{F}} K_4 & A_{\mathscr{F}} \end{bmatrix}\ (i \in \mathscr{F}). \tag{5.18}$$

Theorem 5.3 *If (A, B) is stabilizable, then for any bounded initial states, swarm system (5.1) achieves output containment by dynamic protocols (5.3) and (5.4) with K_4 satisfying that $A_{\mathscr{F}} - B_{\mathscr{F}} K_4$ is Hurwitz, $K_1 = A_{\mathscr{F}} - B_{\mathscr{F}} K_4$, $K_3 = -[\mathrm{Re}(\lambda_1)]^{-1}(R_{\mathscr{F}}^{-1} C_{\mathscr{F}} P_{\mathscr{F}})^T$, and $K_2 = K_3 C_{\mathscr{F}}$, where $P_{\mathscr{F}} = P_{\mathscr{F}}^T > 0$ is the solution to the following algebraic Riccati equation*

$$P_{\mathscr{F}} A_{\mathscr{F}}^T + A_{\mathscr{F}} P_{\mathscr{F}} - P_{\mathscr{F}} C_{\mathscr{F}}^T R_{\mathscr{F}}^{-1} C_{\mathscr{F}} P_{\mathscr{F}} + Q_{\mathscr{F}} = 0, \tag{5.19}$$

for $R_{\mathscr{F}} = R_{\mathscr{F}}^T > 0$ and $Q_{\mathscr{F}} = D_{\mathscr{F}}^T D_{\mathscr{F}} \geq 0$ with $(D_{\mathscr{F}}, A_{\mathscr{F}}^T)$ detectable.

Proof Let $K_1 = A_{\mathscr{F}} - B_{\mathscr{F}} K_4$ and $K_2 = K_3 C_{\mathscr{F}}$, then one can obtain

$$\Upsilon_{\mathscr{F}i} = \begin{bmatrix} A_{\mathscr{F}} - B_{\mathscr{F}} K_4 + \lambda_i K_3 C_{\mathscr{F}} & \lambda_i K_3 C_{\mathscr{F}} \\ B_{\mathscr{F}} K_4 & A_{\mathscr{F}} \end{bmatrix}\ (i \in \mathscr{F}).$$

It can be verified that $\Upsilon_{\mathscr{F}i}$ $(i \in \mathscr{F})$ is similar to

$$\bar{\Upsilon}_{\mathscr{F}i} = \begin{bmatrix} A_{\mathscr{F}} + \lambda_i K_3 C_{\mathscr{F}} & 0 \\ B_{\mathscr{F}} K_4 & A_{\mathscr{F}} - B_{\mathscr{F}} K_4 \end{bmatrix}\ (i \in \mathscr{F}).$$

Since T and $\tilde{T}$ are nonsingular, if (A, B) is stabilizable, then by the PBH criterion for stabilizability, one has

$$\operatorname{rank}\left(\begin{bmatrix} sI - \bar{A}_{11} & -\tilde{E}_1 & 0 & \bar{B}_1 \\ -\tilde{F}_1 & sI - \tilde{D}_1 & 0 & \tilde{B}_1 \\ -\tilde{F}_2 & -\tilde{D}_2 & sI - \tilde{D}_3 & \tilde{B}_2 \end{bmatrix}\right) = n,$$

where $s \in \bar{\mathbb{C}}^+ = \{s | s \in \mathbb{C}, \operatorname{Re}(s) \geq 0\}$. It also can be shown that

$$\operatorname{rank}\left(\begin{bmatrix} sI - \bar{A}_{11} & -\tilde{E}_1 & \bar{B}_1 \\ -\tilde{F}_1 & sI - \tilde{D}_1 & \tilde{B}_1 \end{bmatrix}\right) = q + g \ (\forall s \in \bar{\mathbb{C}}^+),$$

which means that $[sI - A_{\mathscr{F}}, B_{\mathscr{F}}]$ is of full row rank for any $s \in \bar{\mathbb{C}}^+$. Therefore, $(A_{\mathscr{F}}, B_{\mathscr{F}})$ is stabilizable and there exists K_4 such that $A_{\mathscr{F}} - B_{\mathscr{F}} K_4$ is Hurwitz.

Note that $(\tilde{D}_1, \tilde{E}_1)$ is completely observable, one knows that for any $s \in \mathbb{C}$,

$$\begin{bmatrix} sI - \bar{A}_{11} & -\tilde{E}_1 \\ -\tilde{F}_1 & sI - \tilde{D}_1 \\ I & 0 \end{bmatrix}$$

is of full column rank, which means that $(A_{\mathscr{F}}, C_{\mathscr{F}})$ is completely observable. As a result, $(A_{\mathscr{F}}^T, C_{\mathscr{F}}^T)$ is completely controllable. Therefore, for any given $R_{\mathscr{F}} = R_{\mathscr{F}}^T > 0$ and $Q_{\mathscr{F}} = D_{\mathscr{F}}^T D_{\mathscr{F}} \geqslant 0$ with $(D_{\mathscr{F}}, A_{\mathscr{F}}^T)$ detectable, the algebraic Riccati equation (5.19) has a unique solution $P_{\mathscr{F}} = P_{\mathscr{F}}^T > 0$. By Lemma 5.1, one knows that $\operatorname{Re}(\lambda_1) > 0$. Consider the stability of the following subsystem

$$\dot{\bar{\varphi}}_{\mathscr{F}i}(t) = \left(A_{\mathscr{F}}^T + \lambda_i^H C_{\mathscr{F}}^T K_3^T\right) \bar{\varphi}_{\mathscr{F}i}(t) \ (i \in \mathscr{F}). \tag{5.20}$$

Choose the Lyapunov functional candidate as follows:

$$\bar{V}_i(t) = \bar{\varphi}_{\mathscr{F}i}^H(t) P_{\mathscr{F}} \bar{\varphi}_{\mathscr{F}i}(t) \ (i \in \mathscr{F}). \tag{5.21}$$

Let $K_3 = -[\operatorname{Re}(\lambda_1)]^{-1}(R_{\mathscr{F}}^{-1} C_{\mathscr{F}} P_{\mathscr{F}})^T$. Taking the time derivative of $\bar{V}_i(t)$ along the trajectory of (5.20), it holds that

$$\dot{\bar{V}}_i(t) = -\bar{\varphi}_{\mathscr{F}i}^H(t) Q_{\mathscr{F}} \bar{\varphi}_{\mathscr{F}i}(t) + \bar{\varphi}_{\mathscr{F}i}^H(t) \left(1 - 2\operatorname{Re}(\lambda_i)[\operatorname{Re}(\lambda_1)]^{-1}\right) P_{\mathscr{F}} C_{\mathscr{F}}^T R_{\mathscr{F}}^{-1} C_{\mathscr{F}} P_{\mathscr{F}} \bar{\varphi}_{\mathscr{F}i}(t) \leq 0.$$

Note that $(D_{\mathscr{F}}, A_{\mathscr{F}}^T)$ is detectable and $R_{\mathscr{F}} = R_{\mathscr{F}}^T > 0$, one knows that $A_{\mathscr{F}}^T + \lambda_i^H C_{\mathscr{F}}^T K_3^T$ $(i \in \mathscr{F})$ is Hurwitz, which means that $A_{\mathscr{F}} + \lambda_i K_3 C_{\mathscr{F}}$ $(i \in \mathscr{F})$ is Hurwitz. Therefore, $\Upsilon_{\mathscr{F}i}$ $(i \in \mathscr{F})$ is Hurwitz and the observable modes of $(\Upsilon_{\mathscr{F}i}, [0\ I\ 0])$ is asymptotically stable. From Theorem 5.1, the conclusion of Theorem 5.3 can be obtained.

Remark 5.3 From Theorem 5.3, one sees that if (A, B) is stabilizable, then the gain matrices in the dynamic output containment protocol can be determined by solving an algebraic Riccati equation. Although output containment problems have been investigated in [10], only sufficient conditions were obtained, and both a cone complementarity linearization algorithm and an interactive linear matrix inequality algorithm were required to determine the gain matrices in the protocol. The existence of the controller cannot be ensured using the results in [10]. Moreover, in the case that $M = N - 1$, the results in the current paper can be applied to solve the output consensus tracking problems.

5.2.4 Numerical Simulations

In this subsection, a numerical example is given to illustrate the effectiveness of theoretical results obtained in the previous subsections.

Consider a fifth-order swarm system with three leaders and eight followers. The dynamics of each agent is described by (5.1) with $x_i(t)=[x_{i1}(t), x_{i2}(t), \ldots, x_{i5}(t)]^T$, $y_i(t) = [y_{i1}(t), y_{i2}(t), y_{i3}(t)]^T$ $(i = 1, 2, \ldots, 11)$ and

$$A = \begin{bmatrix} -9 & 6 & 7 & -2 & 9 \\ -23 & 10 & 5 & -6 & 23 \\ -2 & 0 & -2 & 0 & -6 \\ -7 & -2 & 1 & -2 & -9 \\ 1 & -2 & 1 & 2 & -5 \end{bmatrix}, \quad B = \begin{bmatrix} 1 \\ 3 \\ 0 \\ 1 \\ -1 \end{bmatrix}, \quad C = \begin{bmatrix} 1 & 0 & 0 & 0 & 1 \\ 1 & 0 & 1 & 0 & -1 \\ 0 & 1 & 0 & -1 & 0 \end{bmatrix}.$$

The interaction topology of the swarm system is shown in Fig. 5.1.

Choose

$$\bar{C} = \begin{bmatrix} 1 & 0 & -1 & 0 & -1 \\ 0 & -1 & 1 & 0 & -1 \end{bmatrix}.$$

It can be shown that $(\bar{A}_{22}, \bar{A}_{12})$ is not completely observable, then choose the following nonsingular matrix $\tilde{T}$ for observability decomposition

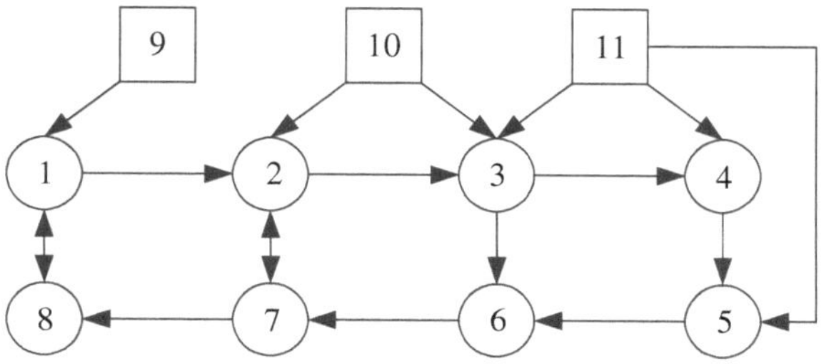

Fig. 5.1 Directed interaction topology G

$$\tilde{T} = \begin{bmatrix} 1 & -1 \\ -1 & 2 \end{bmatrix}.$$

Let $K_4 = [-2.5395, 6.9380, 1.9890, -7.1270]$, then $A_{\mathscr{F}} - B_{\mathscr{F}} K_4$ is Hurwitz. Since (A, B) is stabilizable, by Theorem 5.3, one can obtain that

$$K_1 = \begin{bmatrix} -4 & 4 & 0 & -4 \\ 1.079 & -13.876 & 0.022 & 10.254 \\ 9.079 & -17.876 & 0.022 & 6.254 \\ 13.079 & -9.876 & 0.022 & 10.254 \end{bmatrix}, \; K_2 = \begin{bmatrix} -0.1088 & -0.0750 & -0.3467 & 0 \\ -0.0750 & -0.3460 & -0.6300 & 0 \\ -0.3467 & -0.6300 & -1.8779 & 0 \\ 0.0488 & -0.5132 & -0.6358 & 0 \end{bmatrix},$$

$$K_3 = \begin{bmatrix} -0.1088 & -0.0750 & -0.3467 \\ -0.0750 & -0.3460 & -0.6300 \\ -0.3467 & -0.6300 & -1.8779 \\ 0.0488 & -0.5132 & -0.6358 \end{bmatrix}.$$

Let $z_i(t) = [z_{i1}(t), z_{i2}(t), z_{i3}(t), z_{i4}(t)]^T$ $(i = 1, 2, \ldots, 11)$. Let the initial states of each agent be $x_{ij}(0) = 4(\Theta - 0.5)$ $(i = 1, 2, \ldots, 11;\ j = 1, 2, \ldots, 5)$ and $z_{ij}(0) = 4(\Theta - 0.5)$ $(i = 1, 2, \ldots, 11;\ j = 1, 2, 3, 4)$. Figures 5.2 and 5.3 show the output snapshots of the 11 agents at different time, where the outputs of leaders

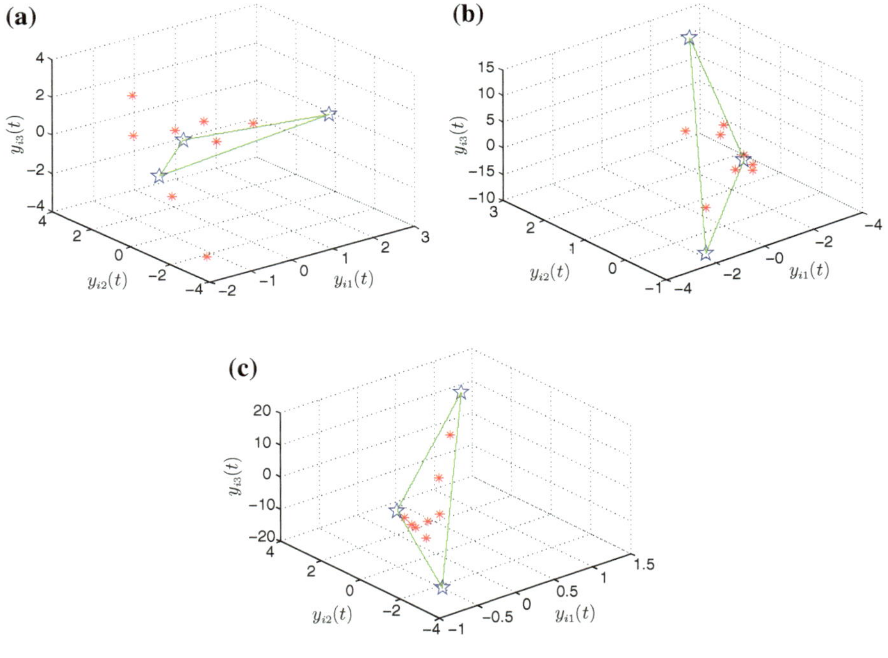

Fig. 5.2 Output snapshots of 11 agents at different time. **a** $t = 0$ s. **b** $t = 5$ s. **c** $t = 50$ s

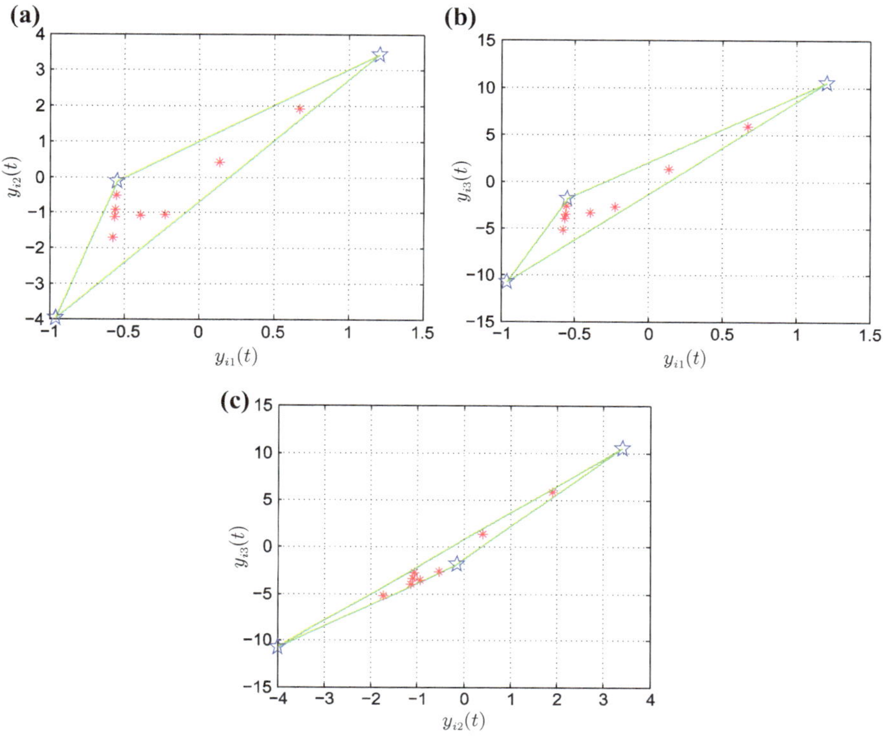

Fig. 5.3 Different views of output snapshots at $t = 50$ s. **a** View of $y_{i1}(t) - y_{i2}(t)$. **b** View of $y_{i1}(t) - y_{i3}(t)$. **c** View of $y_{i2}(t) - y_{i3}(t)$

and followers are denoted by pentagrams and asterisks, respectively, and the convex hull formed by the output of leaders is marked by solid lines. From Figs. 5.2 and 5.3, one sees that the outputs of followers converge to the convex hull formed by those of leaders, which means that the swarm system achieves output containment.

5.3 State Containment Control for Singular Swarm Systems with Time Delays

In this section, containment analysis and design problems for high-order linear time-invariant singular swarm systems with time delays are investigated. To eliminate impulse terms in singular swarm systems and ensure that the singular swarm systems can achieve containment, time-delayed protocols are presented for leaders and followers, respectively. By model transformation, containment problems of singular swarm systems are converted into stability problems of multiple low-dimensional time-delayed systems. In terms of linear matrix inequality, sufficient conditions are presented for time-delayed singular swarm systems to achieve state containment, which are independent of the number of agents. By using the method of changing variables, an approach is provided to determine the gain matrices in the protocols.

5.3.1 Problem Description

Consider a high-order LTI singular swarm system with N agents on a directed interaction topology G. Suppose that each agent with high-order LTI singular dynamics is described by

$$E\dot{x}_i(t) = Ax_i(t) + Bu_i(t) \ (i = 1, 2, \ldots, N), \tag{5.22}$$

where $A \in \mathbb{R}^{n \times n}$, $B \in \mathbb{R}^{n \times m}$, $E \in \mathbb{R}^{n \times n}$ with $\operatorname{rank}(E) = r \leqslant n$, $x_i(t) \in \mathbb{R}^n$ is the state, and $u_i(t) \in \mathbb{R}^m$ is the control input.

Definition 5.3 Singular swarm system (5.22) is said to *achieve state containment* if it is regular and impulse-free, and for any given bounded compatible initial states, and any $k \in \mathscr{F}$, there exist nonnegative constants $\beta_{k,j}$ ($j \in \mathscr{L}$) satisfying $\sum_{j=M+1}^{N} \beta_{k,j} = 1$ such that

$$\lim_{t \to \infty} \left(x_k(t) - \sum_{j=M+1}^{N} \beta_{k,j} x_j(t) \right) = 0. \tag{5.23}$$

Consider the following state containment protocol

$$u_i(t) = K_1 x_i(t) + \bar{u}_i(t) \ (i = 1, 2, \ldots, N), \tag{5.24}$$

where $K_1 \in \mathbb{R}^{m \times n}$ is a constant gain matrix and $\bar{u}_i(t) \in \mathbb{R}^m$ will be designed later. Then under protocol (5.24), singular swarm system (5.22) can be rewritten as

$$E\dot{x}_i(t) = (A + BK_1) x_i(t) + B\bar{u}_i(t) \ (i = 1, 2, \ldots, N). \tag{5.25}$$

Let

$$E = [E_1, E_2] = \begin{bmatrix} E_{11} & E_{12} \\ E_{21} & E_{22} \end{bmatrix},$$

where $E_{11} \in \mathbb{R}^{r\times r}$, $E_{12} \in \mathbb{R}^{r\times(n-r)}$, $E_{21} \in \mathbb{R}^{(n-r)\times r}$, $E_{22} \in \mathbb{R}^{(n-r)\times(n-r)}$. Without loss of generality, it is assumed that E_1 is of full column rank. In this case, there exist $\bar{Q}_S \in \mathbb{R}^{r\times(n-r)}$ and $P_S \in \mathbb{R}^{n\times n}$ such that $E_2 = E_1\bar{Q}_S$ and $P_S E_1 = [I, 0]^T$. Then one has

$$P_S E = \begin{bmatrix} I & \bar{Q}_S \\ 0 & 0 \end{bmatrix}.$$

Define

$$Q_S = \begin{bmatrix} I & -\bar{Q}_S \\ 0 & I \end{bmatrix}.$$

Let $x_i(t) = \left[x_{i1}^T(t), x_{i2}^T(t)\right]^T$, $x_{i1}(t) \in \mathbb{R}^r$, $x_{i2}(t) \in \mathbb{R}^{n-r}$, $Q_S^{-1}x_i(t) = \left[\tilde{x}_{i1}^T(t), \tilde{x}_{i2}^T(t)\right]^T$, $\tilde{x}_{i1}(t) \in \mathbb{R}^r$, and $\tilde{x}_{i2}(t) \in \mathbb{R}^{n-r}$. It can be shown that $\tilde{x}_{i1}(t) = x_{i1}(t) + \bar{Q}_S x_{i2}(t)$, $\tilde{x}_{i2}(t) = x_{i2}(t)$, and

$$P_S E Q_S = \begin{bmatrix} I & 0 \\ 0 & 0 \end{bmatrix}.$$

Define

$$P_S\left(A + BK_1\right) Q_S = \begin{bmatrix} \tilde{A}_{11} & \tilde{A}_{12} \\ \tilde{A}_{21} & \tilde{A}_{22} \end{bmatrix}, \quad P_S B = \begin{bmatrix} \tilde{B}_{S1} \\ \tilde{B}_{S2} \end{bmatrix},$$

where $\tilde{A}_{11} \in \mathbb{R}^{r\times r}$, $\tilde{A}_{12} \in \mathbb{R}^{r\times(n-r)}$, $\tilde{A}_{21} \in \mathbb{R}^{(n-r)\times r}$, $\tilde{A}_{22} \in \mathbb{R}^{(n-r)\times(n-r)}$, $\tilde{B}_{S1} \in \mathbb{R}^{r\times m}$, and $\tilde{B}_{S2} \in \mathbb{R}^{(n-r)\times m}$. Then singular swarm system (5.25) can be transformed into

$$\begin{cases} \dot{\tilde{x}}_{i1}(t) = \tilde{A}_{11}\tilde{x}_{i1}(t) + \tilde{A}_{12}\tilde{x}_{i2}(t) + \tilde{B}_{S1}\bar{u}_i(t), \\ \quad 0 = \tilde{A}_{21}\tilde{x}_{i1}(t) + \tilde{A}_{22}\tilde{x}_{i2}(t) + \tilde{B}_{S2}\bar{u}_i(t). \end{cases} \tag{5.26}$$

Assume that $(E, A + BK_1)$ is regular and impulse-free. Then $\bar{A}_{22}$ is nonsingular. In this case, (5.26) can be rewritten as

$$\begin{cases} \dot{\tilde{x}}_{i1}(t) = \left(\tilde{A}_{11} - \tilde{A}_{12}\tilde{A}_{22}^{-1}\tilde{A}_{21}\right)\tilde{x}_{i1}(t) + \left(\tilde{B}_{S1} - \tilde{A}_{12}\tilde{A}_{22}^{-1}\tilde{B}_{S2}\right)\bar{u}_i(t), \\ \tilde{x}_{i2}(t) = -\tilde{A}_{22}^{-1}\tilde{A}_{21}\tilde{x}_{i1}(t) - \tilde{A}_{22}^{-1}\tilde{B}_{S2}\bar{u}_i(t). \end{cases} \tag{5.27}$$

It should be mentioned that if (E, A) is not impulse-free but (E, A, B) is impulse controllable, then there exists K_1 which can eliminate impulse terms in (E, A).

Construct time-delayed $\bar{u}_i(t)$, for all $i = 1, 2, \ldots, N$ as follows:

$$\begin{aligned} \bar{u}_i(t) = {} & K_2\textstyle\sum_{j\in N_i} w_{ij}\left(\left(x_{i1}(t-\tau) + \bar{Q}_S x_{i2}(t-\tau)\right) - \left(x_{j1}(t-\tau) + \bar{Q}_S x_{j2}(t-\tau)\right)\right) \\ & + K_3\left(x_{i1}(t-\tau) + \bar{Q}_S x_{i2}(t-\tau)\right), \ \forall i \in \mathscr{F}, \end{aligned} \tag{5.28}$$

$$\bar{u}_i(t) = K_3\left(x_{i1}(t-\tau) + \bar{Q}_S x_{i2}(t-\tau)\right), \ \forall i \in \mathscr{L}, \tag{5.29}$$

where K_2, $K_3 \in \mathbb{R}^{m \times q}$ are constant gain matrices, and $\tau > 0$ is assumed to be a known constant time delay.

Remark 5.4 In the time-delayed protocols (5.24), (5.28), and (5.29), K_1 is used to eliminate impulse terms of the states of singular system (5.22), K_3 can be used to assign the motion modes of leaders, and K_2 is designed to ensure that the states of the followers can converge to the convex hull formed by the states of the leaders. In [8, 9, 11], it was assumed that the states of leaders were only dependent on the initial states of the leaders, and the protocols were only designed for followers. However, in some practical cases, the motion modes of leaders may need to be assigned. In these cases, the protocols in [8, 9, 11] are not valid. Moreover, the gain matrices K_1, K_2, and K_3 should be designed in the following sequence. First, design K_1 to eliminate impulse terms in singular system (5.22). Then design K_3 to specify the motion modes of the leaders. Finally, K_2 is designed to guarantee that the singular swarm system can achieve containment. Design approaches for K_1 can be found in [12].

The main objective of this section is to investigate the following two main problems for singular swarm system (5.22) with protocols (5.24), (5.28), and (5.29): (i) under what conditions state containment can be achieved; (ii) how to determine the gain matrices in the protocols.

5.3.2 Problem Transformation and Preliminary Results

In this subsection, containment problems of singular swarm system (5.22) with protocols (5.24), (5.28), and (5.29) are transformed into the stability problems of M time-delayed systems, and preliminary results are presented. Let

$$x_{\mathscr{F}1}(t) = \left[x_{11}^T(t), x_{21}^T(t), \ldots, x_{M1}^T(t)\right]^T, x_{\mathscr{L}1}(t) = \left[x_{(M+1)1}^T(t), x_{(M+2)1}^T(t), \ldots, x_{N1}^T(t)\right]^T,$$

$$x_{\mathscr{F}2}(t) = \left[x_{12}^T(t), x_{22}^T(t), \ldots, x_{M2}^T(t)\right]^T, x_{\mathscr{L}2}(t) = \left[x_{(M+1)2}^T(t), x_{(M+2)2}^T(t), \ldots, x_{N2}^T(t)\right]^T,$$

$$\tilde{x}_{\mathscr{F}1}(t) = \left[\tilde{x}_{11}^T(t), \tilde{x}_{21}^T(t), \ldots, \tilde{x}_{M1}^T(t)\right]^T, \tilde{x}_{\mathscr{L}1}(t) = \left[\tilde{x}_{(M+1)1}^T(t), \tilde{x}_{(M+2)1}^T(t), \ldots, \tilde{x}_{N1}^T(t)\right]^T,$$

$$\tilde{x}_{\mathscr{F}2}(t) = \left[\tilde{x}_{12}^T(t), \tilde{x}_{22}^T(t), \ldots, \tilde{x}_{M2}^T(t)\right]^T, \tilde{x}_{\mathscr{L}2}(t) = \left[\tilde{x}_{(M+1)2}^T(t), \tilde{x}_{(M+2)2}^T(t), \ldots, \tilde{x}_{N2}^T(t)\right]^T.$$

Under protocols (5.24), (5.28), and (5.29), the dynamics of the singular swarm system (5.22) can be described in a compact form as

$$\begin{cases} \dot{\tilde{x}}_{\mathscr{F}1}(t) = \left(I \otimes \left(\tilde{A}_{11} - \tilde{A}_{12}\tilde{A}_{22}^{-1}\tilde{A}_{21}\right)\right)\tilde{x}_{\mathscr{F}1}(t) + L_2 \otimes \left(\tilde{B}_{S1} - \tilde{A}_{12}\tilde{A}_{22}^{-1}\tilde{B}_{S2}\right)K_2\tilde{x}_{\mathscr{L}1}(t-\tau) \\ \quad + \left(L_1 \otimes \left(\tilde{B}_{S1} - \tilde{A}_{12}\tilde{A}_{22}^{-1}\tilde{B}_{S2}\right)K_2 + I \otimes \left(\tilde{B}_{S1} - \tilde{A}_{12}\tilde{A}_{22}^{-1}\tilde{B}_{S2}\right)K_3\right)\tilde{x}_{\mathscr{F}1}(t-\tau),\ t > 0, \\ \tilde{x}_{\mathscr{F}1}(t) = \varphi_{\mathscr{F}1}(t),\ t \in [-\tau, 0], \end{cases} \tag{5.30}$$

$$\begin{aligned} \tilde{x}_{\mathscr{F}2}(t) = {} & I \otimes \left(-\tilde{A}_{22}^{-1}\tilde{A}_{21}\right)\tilde{x}_{\mathscr{F}1}(t) + L_2 \otimes \left(-\tilde{A}_{22}^{-1}\tilde{B}_{S2}K_2\right)\tilde{x}_{\mathscr{L}1}(t-\tau) \\ & + \left(L_1 \otimes \left(-\tilde{A}_{22}^{-1}\tilde{B}_{S2}K_2\right) + I \otimes \left(-\tilde{A}_{22}^{-1}\tilde{B}_{S2}K_3\right)\right)\tilde{x}_{\mathscr{F}1}(t-\tau), \end{aligned} \tag{5.31}$$

$$\begin{cases} \dot{\tilde{x}}_{\mathscr{L}1}(t) = \left(I \otimes \left(\tilde{A}_{11} - \tilde{A}_{12}\tilde{A}_{22}^{-1}\tilde{A}_{21}\right)\right)\tilde{x}_{\mathscr{L}1}(t) \\ \quad + \left(I \otimes \left(\tilde{B}_{S1} - \tilde{A}_{12}\tilde{A}_{22}^{-1}\tilde{B}_{S2}\right)K_3\right)\tilde{x}_{\mathscr{L}1}(t-\tau),\ t > 0, \\ \tilde{x}_{\mathscr{L}1}(t) = \varphi_{\mathscr{L}1}(t),\ t \in [-\tau, 0], \end{cases} \tag{5.32}$$

$$\tilde{x}_{\mathscr{L}2}(t) = I \otimes \left(-\tilde{A}_{22}^{-1}\tilde{A}_{21}\right)\tilde{x}_{\mathscr{L}1}(t) + I \otimes \left(-\tilde{A}_{22}^{-1}\tilde{B}_{S2}K_3\right)\tilde{x}_{\mathscr{L}1}(t-\tau), \tag{5.33}$$

where $\varphi_{\mathscr{F}1}(t)$ and $\varphi_{\mathscr{L}1}(t)$ are bounded compatible continuous vector-valued functions on $t \in [-\tau, 0]$.

Remark 5.5 From (5.32), one sees that the motion modes of the leaders can be partially assigned by choosing state feedback gain matrix K_3 for system $\left(\tilde{A}_{11} - \tilde{A}_{12}\tilde{A}_{22}^{-1}\tilde{A}_{21},\ \tilde{B}_{S1} - \tilde{A}_{12}\tilde{A}_{22}^{-1}\tilde{B}_{S2}\right)$ with a constant input delay. The control for such systems has been extensively studied in the literature (see, e.g., [13–16] and references therein).

Let

$$\theta_{Si}(t) = \sum_{j_1 \in N_{i1}} w_{ij}(\tilde{x}_{i1}(t-\tau) - \tilde{x}_{j1}(t-\tau)),\ \forall i \in \mathscr{F},$$

and $\theta_S(t) = [\theta_{S1}^T(t), \theta_{S2}^T(t), \ldots, \theta_{SM}^T(t)]^T$, then one can obtain

$$\theta_S(t) = (L_2 \otimes I)\tilde{x}_{\mathscr{L}1}(t-\tau) + (L_1 \otimes I)\tilde{x}_{\mathscr{F}1}(t-\tau). \tag{5.34}$$

Taking the derivative of (5.34) with respect to t, one has

$$\dot{\theta}_S(t) = (L_2 \otimes I)\dot{\tilde{x}}_{\mathscr{L}1}(t-\tau) + (L_1 \otimes I)\dot{\tilde{x}}_{\mathscr{F}1}(t-\tau). \tag{5.35}$$

Substituting (5.30), (5.32), and (5.34) into (5.35), one can obtain

$$\begin{aligned} \dot{\theta}_S(t) = {} & I \otimes \left(\tilde{A}_{11} - \tilde{A}_{12}\tilde{A}_{22}^{-1}\tilde{A}_{21}\right)\theta_S(t) + \left(L_1 \otimes \left(\tilde{B}_{S1} - \tilde{A}_{12}\tilde{A}_{22}^{-1}\tilde{B}_{S2}\right)K_2 \right. \\ & \left. + I \otimes \left(\tilde{B}_{S1} - \tilde{A}_{12}\tilde{A}_{22}^{-1}\tilde{B}_{S2}\right)K_3\right)\theta_S(t-\tau). \end{aligned} \tag{5.36}$$

It can be seen that if system (5.36) is asymptotically stable, then $\lim_{t\to\infty}\theta_S(t)=0$; that is,

$$\lim_{t\to\infty}\left(\tilde{x}_{\mathscr{F}1}(t-\tau)-\left(-L_1^{-1}L_2\otimes I\right)\tilde{x}_{\mathscr{L}1}(t-\tau)\right)=0,$$

which means that

$$\lim_{t\to\infty}\left(\tilde{x}_{\mathscr{F}1}(t)-\left(-L_1^{-1}L_2\otimes I\right)\tilde{x}_{\mathscr{L}1}(t)\right)=0,\ \forall i\in\mathscr{F}. \tag{5.37}$$

From (5.31), (5.33), and (5.37), one has

$$\lim_{t\to\infty}\left(\tilde{x}_{\mathscr{F}2}(t)-\left(-L_1^{-1}L_2\otimes I\right)\tilde{x}_{\mathscr{L}2}(t)\right)=0,$$

Due to $\tilde{x}_{i2}(t)=x_{i2}(t)$, the following holds

$$\lim_{t\to\infty}\left(x_{\mathscr{F}2}(t)-\left(-L_1^{-1}L_2\otimes I\right)x_{\mathscr{L}2}(t)\right)=0. \tag{5.38}$$

Note that

$$x_{\mathscr{F}1}(t)=\tilde{x}_{\mathscr{F}1}(t)-\left(I\otimes\bar{Q}_S\right)x_{\mathscr{F}2}(t). \tag{5.39}$$

From (5.37) to (5.39), one can obtain

$$\lim_{t\to\infty}\left(x_{\mathscr{F}1}(t)-\left(-L_1^{-1}L_2\otimes I\right)x_{\mathscr{L}1}(t)\right)=0. \tag{5.40}$$

From Lemma 5.1, one can see that all the states of followers converge to the convex hull formed by the states of the leaders; that is, singular swarm system (5.22) with protocols (5.24), (5.28), and (5.29) achieves containment.

Under Assumption 1, one knows that all the eigenvalues of L_1 have positive real parts. Let $\tilde{\theta}_S(t)=[\eta_{S1}^T(t),\eta_{S2}^T(t),\ldots,\eta_{SM}^T(t)]^T=(U_{\mathscr{F}}^{-1}\otimes I)\theta_S(t)$, where $U_{\mathscr{F}}$ is the same with the one defined in Sect. 5.2. From (5.36), one has

$$\dot{\tilde{\theta}}_S(t)=\left(I\otimes A_\eta\right)\tilde{\theta}_S(t)+\left(\bar{\Lambda}_{\mathscr{F}}\otimes B_\eta K_2+I\otimes B_\eta K_3\right)\tilde{\theta}_S(t-\tau), \tag{5.41}$$

where $A_\eta=\tilde{A}_{11}-\tilde{A}_{12}\tilde{A}_{22}^{-1}\tilde{A}_{21}$ and $B_\eta=\tilde{B}_{S1}-\tilde{A}_{12}\tilde{A}_{22}^{-1}\tilde{B}_{S2}$. Since $\bar{\Lambda}_{\mathscr{F}}$ is an upper-triangular matrix, the following lemma can be easily obtained.

Lemma 5.2 *Singular swarm system (5.22) with protocols (5.24), (5.28), and (5.29) achieves containment if $(E, A+BK_1)$ is regular and impulse-free, and the following system*

$$\dot{\eta}_{Si}(t)=A_\eta\eta_{Si}(t)+B_\eta\left(\lambda_i K_2+K_3\right)\eta_{Si}(t-\tau),\ \forall i\in\mathscr{F}, \tag{5.42}$$

is asymptotically stable.

From (5.26), one knows that

$$\begin{bmatrix} I & 0 \\ 0 & 0 \end{bmatrix} \begin{bmatrix} \dot{\tilde{x}}_{i1}(t) \\ \dot{\tilde{x}}_{i2}(t) \end{bmatrix} = \begin{bmatrix} \tilde{A}_{11} & \tilde{A}_{12} \\ \tilde{A}_{21} & \tilde{A}_{22} \end{bmatrix} \begin{bmatrix} \tilde{x}_{i1}(t) \\ \tilde{x}_{i2}(t) \end{bmatrix} + \begin{bmatrix} \tilde{B}_{S1} \\ \tilde{B}_{S2} \end{bmatrix} \bar{u}_i(t).$$

Assume that $(E, A + BK_1)$ is impulse-free. Let

$$\bar{P}_S = \begin{bmatrix} I & -\tilde{A}_{12}\tilde{A}_{22}^{-1} \\ 0 & I \end{bmatrix}, \ \bar{Q}_S = \begin{bmatrix} I & 0 \\ -\tilde{A}_{22}^{-1}\tilde{A}_{21} & \tilde{A}_{22}^{-1} \end{bmatrix}, \ \tilde{x}(t) = \bar{Q}_S \bar{x}(t).$$

One has

$$\begin{bmatrix} I & 0 \\ 0 & 0 \end{bmatrix} \bar{Q}_S \begin{bmatrix} \dot{\bar{x}}_{i1}(t) \\ \dot{\bar{x}}_{i2}(t) \end{bmatrix} = \begin{bmatrix} \tilde{A}_{11} & \tilde{A}_{12} \\ \tilde{A}_{21} & \tilde{A}_{22} \end{bmatrix} \bar{Q}_S \begin{bmatrix} \bar{x}_{i1}(t) \\ \bar{x}_{i2}(t) \end{bmatrix} + \begin{bmatrix} \tilde{B}_{S1} \\ \tilde{B}_{S2} \end{bmatrix} \bar{u}_i(t). \quad (5.43)$$

Pre-multiplying the left and right sides of (5.43) by $\bar{P}_S$, one can obtain

$$\begin{bmatrix} I & 0 \\ 0 & 0 \end{bmatrix} \dot{\bar{x}} = \begin{bmatrix} A_\eta & 0 \\ 0 & I \end{bmatrix} \bar{x}(t) + \begin{bmatrix} B_\eta \\ \tilde{B}_{S2} \end{bmatrix} \bar{u}_i(t). \quad (5.44)$$

By Lemma 2.16, the following lemma can be obtained.

Lemma 5.3 *(A_η, B_η) is controllable or stabilizable if and only if $(E, A + BK_1)$ is regular and impulse-free, and $(E, A + BK_1, B)$ is $\mathbb{R}$-controllable or stabilizable.*

5.3.3 State Containment Analysis and Protocol Design

In this subsection, first, based on LMI techniques, sufficient conditions for singular swarm system (5.22) with protocols (5.24), (5.28), and (5.29) to achieve containment are presented, which include only four LMI constraints independent of the number of agents. Then by using the method of changing variables, design approaches for the gain matrices in the protocols are given.

Lemma 5.4 ([17]) *Let $\eta_S(t) \in \mathbb{R}^r$ be a vector-valued function with first-order continuous-derivative entries. Then the following integral inequality holds for any matrices $M_1, M_2 \in \mathbb{R}^{r\times r}$ and $S_S = S_S^T > 0$*

$$-\int_{t-\tau}^{t} \dot{\eta}_S^T(s) S_S \dot{\eta}_S(s) ds \le \varsigma_S^T(t) \begin{bmatrix} M_1^T + M_1 & -M_1^T + M_2 \\ * & -M_2^T - M_2 \end{bmatrix} \times \varsigma_S(t) + \tau \varsigma_S^T(t)$$

$$\times \begin{bmatrix} M_1^T \\ M_2^T \end{bmatrix} S_S^{-1} [M_1, M_2] \varsigma_S(t),$$

where $\varsigma_S(t) = \left[\eta_S^T(t), \eta_S^T(t-\tau)\right]^T$.

Theorem 5.4 *For any given bounded compatible initial states, singular swarm system (5.22) with protocols (5.24), (5.28), and (5.29) achieves containment if* $(E, A + BK_1)$ *is regular and impulse-free, and there exist* $2r \times 2r$ *real matrices* $R_S = R_S^T > 0$, $\Theta_S = \Theta_S^T > 0$, $S_S = S_S^T > 0$, M_1, *and* M_2, *such that the following LMIs are feasible:*

$$\bar{\Phi}_i = \begin{bmatrix} \bar{\Phi}_{11} & \bar{\Phi}_{i12} & \tau \Lambda_{A_\eta}^T S_S & \tau M_1^T \\ * & \bar{\Phi}_{22} & \tau \left(\Psi_{\bar{\lambda}_i} \Lambda_{B_\eta} \Lambda_{K_2} + \Lambda_{B_\eta} \Lambda_{K_3} \right)^T S_S & \tau M_2^T \\ * & * & -\tau S_S & 0 \\ * & * & * & -\tau S_S \end{bmatrix} < 0 \; (i = 1, 2, 3, 4), \quad (5.45)$$

where

$$\bar{\Phi}_{11} = \Lambda_{A_\eta}^T R_S + R_S \Lambda_{A_\eta} + \Theta_S + M_1^T + M_1,$$

$$\bar{\Phi}_{i12} = R_S \left(\Psi_{\bar{\lambda}_i} \Lambda_{B_\eta} \Lambda_{K_2} + \Lambda_{B_\eta} \Lambda_{K_3} \right) - M_1^T + M_2,$$

$$\bar{\Phi}_{22} = -\Theta_S - M_2^T - M_2.$$

Proof Consider the stability of the following system:

$$\dot{\eta}_{Si}(t) = A_\eta \eta_{Si}(t) + B_\eta \left(\lambda_i K_2 + K_3 \right) \eta_{Si}(t - \tau), i \in \mathscr{L}. \quad (5.46)$$

By the decomposition of real and imaginary parts, it can be shown that asymptotic stability of system (5.46) is equivalent to that of the following system:

$$\dot{\hat{\eta}}_{Si}(t) = \Lambda_{A_\eta} \hat{\eta}_{Si}(t) + \left(\Psi_{\lambda_i} \Lambda_{B_\eta} \Lambda_{K_2} + \Lambda_{B_\eta} \Lambda_{K_3} \right) \hat{\eta}_{Si}(t - \tau). \quad (5.47)$$

Consider a Lyapunov–Krasovskii functional candidate as follows:

$$V_i(t) = V_{i1}(t) + V_{i2}(t) + V_{i3}(t), \quad (5.48)$$

where

$$V_{i1}(t) = \hat{\eta}_{Si}^T(t) R_S \hat{\eta}_{Si}(t),$$

$$V_{i2}(t) = \int_{t-\tau}^{t} \hat{\eta}_{Si}^T(s) \Theta_S \hat{\eta}_{Si}(s) ds,$$

$$V_{i3}(t) = \int_{-\tau}^{0} \int_{t+\sigma}^{t} \dot{\hat{\eta}}_i^T(s) S_S \dot{\hat{\eta}}_i(s) ds d\sigma.$$

Taking the time derivative of $V_{ij}(t)$ $(j = 1, 2, 3)$ along the trajectory of (5.47), one has

$$\begin{aligned} \dot{V}_{i1}(t) &= 2\hat{\eta}_{Si}^T(t) R_S H_{Si} \varsigma_i(t), \\ \dot{V}_{i2}(t) &= \hat{\eta}_{Si}^T(t)\Theta_S\hat{\eta}_{Si}(t) - \hat{\eta}_{Si}^T(t-\tau)\Theta_S\hat{\eta}_{Si}(t-\tau), \\ \dot{V}_{i3}(t) &= \tau\varsigma_{Si}^T(t)H_{Si}^T S H_{Si}\varsigma_{Si}(t) - \int_{t-\tau}^t \dot{\hat{\eta}}_i^T(s)S_S\dot{\hat{\eta}}_i(s)ds, \end{aligned} \tag{5.49}$$

where $H_{Si} = [\Lambda_{A_\eta}, \Psi_{\lambda_i}\Lambda_{B_\eta}\Lambda_{K_2} + \Lambda_{B_\eta}\Lambda_{K_3}]$ and $\varsigma_{Si}(t) = [\hat{\eta}_{Si}^T(t), \hat{\eta}_{Si}^T(t-\tau)]^T$. By Lemma 5.4, for any $M_1, M_2 \in \mathbb{R}^{2r\times 2r}$ one has

$$\begin{aligned} -\int_{t-\tau}^t \dot{\hat{\eta}}_{Si}^T(s)S_S\dot{\hat{\eta}}_{Si}(s)ds \le \varsigma_{Si}^T(t)\begin{bmatrix} M_1^T + M_1 & -M_1^T + M_2 \\ * & -M_2^T - M_2 \end{bmatrix}\varsigma_{Si}(t) + \tau\varsigma_{Si}^T(t) \\ \times\begin{bmatrix} M_1^T \\ M_2^T \end{bmatrix} S_S^{-1}[M_1, M_2]\varsigma_{Si}(t). \end{aligned} \tag{5.50}$$

From (5.48) to (5.50), one can obtain that

$$\dot{V}_i(t) \le \varsigma_{Si}^T(t)\left(\bar{\Phi}_{Si} + \tau H_{Si}^T S_S H_{Si} + \tau\begin{bmatrix} M_1^T \\ M_2^T \end{bmatrix} S_S^{-1}[M_1, M_2]\right)\varsigma_{Si}(t),$$

where

$$\bar{\Phi}_{Si} = \begin{bmatrix} \bar{\Phi}_{11} & R_S\left(\Psi_{\lambda_i}\Lambda_{B_\eta}\Lambda_{K_2} + \Lambda_{B_\eta}\Lambda_{K_3}\right) - M_1^T + M_2 \\ * & \bar{\Phi}_{22} \end{bmatrix}.$$

By Schur complement in Lemma 2.4, if $\bar{\Phi}_i < 0$ with $\bar{\lambda}_i = \lambda_i$, for all $i \in \mathscr{F}$ are feasible, system (5.47) is asymptotically stable. Then from Lemmas 5.3 and 4.3, singular swarm system (5.22) with protocols (5.24), (5.28), and (5.29) achieves containment if $(E, A + BK_1)$ is regular and impulse-free, and $\bar{\Phi}_i < 0$, for all $i = 1, 2, 3, 4$ are feasible. The proof of Theorem 5.4 is completed.

If the gain matrix K_2 is unknown, then $\bar{\Phi}_i < 0$ $(i = 1, 2, 3, 4)$ in Theorem 5.4 becomes nonlinear matrix inequalities. In this case, it is difficult to solve these inequalities for K_2. In the following, by using the method of changing variable, an approach to determine K_2 is proposed.

Theorem 5.5 *For any given bounded compatible initial states, singular swarm system (5.22) with protocols (5.24), (5.28), and (5.29) can achieve containment if $(E, A + BK_1)$ is regular and impulse-free, and there exist positive symmetric matrices $\bar{R}_S \in \mathbb{R}^{2r\times 2r}$, $\bar{S}_S \in \mathbb{R}^{2r\times 2r}$ $\bar{\Theta}_S \in \mathbb{R}^{r\times r}$, and matrix $\bar{K}_2 \in \mathbb{R}^{m\times r}$, such that the following LMIs are feasible*

$$\tilde{\Phi}_i = \begin{bmatrix} \tilde{\Phi}_{i11} & \tilde{\Phi}_{i12} & \tilde{\Phi}_{i13} & 0 & \bar{R}_S \\ * & \tilde{\Phi}_{22} & \tilde{\Phi}_{i23} & \bar{S}_S & 0 \\ * & * & -\tau^{-1}\bar{S}_S & 0 & 0 \\ * & * & * & -\tau^{-1}\bar{S}_S & 0 \\ * & * & * & * & -\Lambda_{\bar{\Theta}_S} \end{bmatrix} < 0\,(i = 1, 2, 3, 4), \quad (5.51)$$

where

$$\tilde{\Phi}_{i11} = \Lambda_{A_\eta}\bar{R}_S + \bar{R}_S\Lambda_{A_\eta}^T + \Psi_{\bar{\lambda}_i}\Lambda_{B_\eta}\Lambda_{\bar{K}_2} + \Lambda_{\bar{K}_2}^T\Lambda_{B_\eta}^T\Psi_{\bar{\lambda}_i}^T + \Lambda_{B_\eta}\Lambda_{K_3}\Lambda_{\bar{\Theta}_S} + \Lambda_{\bar{\Theta}_S}^T\Lambda_{K_3}^T\Lambda_{B_\eta}^T - \Lambda_{\bar{\Theta}_S},$$

$$\tilde{\Phi}_{i12} = \bar{R}_S + \Psi_{\bar{\lambda}_i}\Lambda_{B_\eta}\Lambda_{\bar{K}_2} + \Lambda_{B_\eta}\Lambda_{K_3}\Lambda_{\bar{\Theta}_S} - 2\Lambda_{\bar{\Theta}_S},$$

$$\tilde{\Phi}_{i13} = \bar{R}_S\Lambda_{A_\eta}^T + \Lambda_{\bar{K}_2}^T\Lambda_{B_\eta}^T\Psi_{\bar{\lambda}_i}^T + \Lambda_{\bar{\Theta}_S}^T\Lambda_{K_3}^T\Lambda_{B_\eta}^T,$$

$$\tilde{\Phi}_{22} = -3\Lambda_{\bar{\Theta}_S},$$

$$\tilde{\Phi}_{i23} = \Lambda_{\bar{K}_2}^T\Lambda_{B_\eta}^T\Psi_{\bar{\lambda}_i}^T + \Lambda_{\bar{\Theta}_S}^T\Lambda_{K_3}^T\Lambda_{B_\eta}^T.$$

In this case, the gain matrix K_2 *is given by* $K_2 = \bar{K}_2\bar{\Theta}_S^{-1}$.

Proof Let

$$\bar{A}_{\eta i} = \begin{bmatrix} \Lambda_{A_\eta} & \Psi_{\bar{\lambda}_i}\Lambda_{B_\eta}\Lambda_{K_2} + \Lambda_{B_\eta}\Lambda_{K_3} \\ I & -I \end{bmatrix} (i = 1, 2, 3, 4),\ T_S = \begin{bmatrix} R_S & 0 \\ M_1 & M_2 \end{bmatrix},\ M^T = \begin{bmatrix} M_1^T \\ M_2^T \end{bmatrix}.$$

Then, by Schur complement, $\bar{\Phi}_i < 0$ is equivalent to

$$\hat{\Phi}_{Si} = \begin{bmatrix} T_S^T\bar{A}_{\eta i} + \bar{A}_{\eta i}^T T_S + \mathrm{diag}\{\Theta_S, -\Theta_S\} & H_{Si}^T & M^T \\ * & -\tau^{-1}S_S^{-1} & 0 \\ * & * & -\tau^{-1}S_S \end{bmatrix} < 0. \quad (5.52)$$

Consider the case that $M_1 = -R_S$ and $M_2 = \Theta_S$. Thus, one can obtain that

$$T_S^{-1} = \begin{bmatrix} R_S^{-1} & 0 \\ \Theta_S^{-1} & \Theta_S^{-1} \end{bmatrix}. \quad (5.53)$$

Direct computation shows that

$$\begin{aligned}&\operatorname{diag}\left\{T_S^{-T}, I, S_S^{-1}\right\} \hat{\Phi}_{Si} \operatorname{diag}\left\{T_S^{-1}, I, S_S^{-1}\right\} \\ &= \begin{bmatrix} \bar{A}_{\eta i} T_S^{-1} + T_S^{-T} \bar{A}_{\eta i}^T + T_S^{-T} \operatorname{diag}\{\Theta_S, -\Theta_S\} T_S^{-1} & T_S^{-T} H_{Si}^T & \left[0, S_S^{-1}\right]^T \\ * & -\tau^{-1} S_S^{-1} & 0 \\ * & * & -\tau^{-1} S^{-1} \end{bmatrix}.\end{aligned} \tag{5.54}$$

Let $\bar{R}_S = R_S^{-1}$, $\bar{S} = S_S^{-1}$, $\Lambda_{\bar{\Theta}_S} = \Theta_S^{-1}$ and $\bar{K}_2 = K_2 \bar{\Theta}_S$. Then from (5.52) to (5.54), one can obtain that if $\tilde{\Phi}_i < 0$ is feasible, then $\hat{\Phi}_{Si} < 0$. Hence, if $\tilde{\Phi}_i < 0$ $(i = 1, 2, 3, 4)$ are feasible, and $(E, A + BK_1)$ is regular and impulse-free, then singular swarm system (5.22) with protocols (5.24), (5.28), and (5.29) can achieve containment with $K_2 = \bar{K}_2 \bar{\Theta}_S^{-1}$. The proof of Theorem 5.5 is completed.

Remark 5.6 Theorem 5.4 presents sufficient conditions for singular swarm system (5.22) with protocols (5.24), (5.28), and (5.29) to achieve containment, where the integral inequality method shown in Lemma 5.4 is applied to analyze the influence of the time delay. In dealing with the delay-dependent stability problems of time-delayed systems, Zhang et al. [17] showed that this method can largely reduce the conservation of the stability conditions. To further reduce the conservation, we recommend applying the delay partition method shown in [18, 19]. By partitioning the time delay into several components, this method can lead to much less conservative conditions. Moreover, for the case where gain matrix K_2 is unknown, the conditions for stability of systems (5.42) become nonlinear. In this case, the method of changing variable is used to determine gain matrix K_2 in Theorem 5.5, which may also bring in some conservation.

Remark 5.7 The presented results can be applied to solve the containment problems or the consensus tracking problems for normal high-order LTI time-delayed swarm systems by setting $E = I$ or $E = I$ and $N - M = 1$, respectively. In [9] and [11], containment problems and consensus tracking problems for normal high-order LTI swarm systems were investigated, respectively. But they did not consider the time delays and the motion modes of leaders in these papers cannot be assigned.

5.3.4 Numerical Simulations

Consider a fourth-order singular swarm system with 12 agents. Assume that there are four leaders and eight followers. The interaction topology of the swarm system is shown in Fig. 5.4. The dynamics of each agent is described by (5.22) with $i = 1, 2, \ldots, 12$, $x_i(t) = [x_{i11}(t), x_{i12}(t), x_{i21}(t), x_{i22}(t)]^T$ and

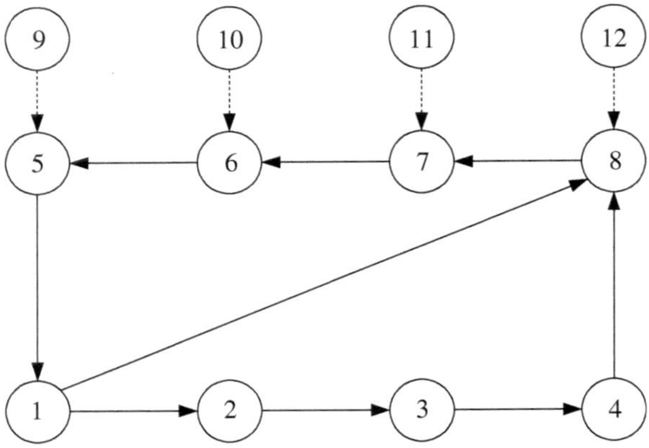

Fig. 5.4 Directed interaction topology G

$$E = \begin{bmatrix} 1 & 0 & \vdots & 1 & 0 \\ 2 & 3 & \vdots & 2 & -6 \\ 5 & 0 & \vdots & 5 & 0 \\ 0 & -6 & \vdots & 0 & 12 \end{bmatrix}, \quad A = \begin{bmatrix} 1 & 1 & -1 & 2 \\ -2 & 5 & 1 & 4 \\ 5 & 5 & -5 & 10 \\ 2 & 1 & 4 & -3 \end{bmatrix}, \quad B = \begin{bmatrix} 0 \\ 0 \\ 1 \\ 0 \end{bmatrix}.$$

One can obtain that E_1 is of full column rank and

$$\bar{Q}_S = \begin{bmatrix} 1 & 0 \\ 0 & -2 \end{bmatrix}, \quad P_S = \begin{bmatrix} 1 & 0 & 0 & 0 \\ -\frac{2}{3} & \frac{1}{3} & 0 & 0 \\ -5 & 0 & 1 & 0 \\ -4 & 2 & 0 & 1 \end{bmatrix}.$$

It can be shown that (E, A) is not impulse-free, but (E, A, B) is impulse controllable. With the design approaches in [12], one can obtain a K_1 to make $(E, A + BK_1)$ impulse-free as $K_1 = [-3, -3, -0.1, 3]$. Set $K_3 = [1.6843, 4.4769]$ to assign the motion modes of the leaders. Let $\tau = 0.025s$. By using the Feasp solver in Matlab's LMI toolbox to solve the LMIs in Theorem 5.5, one can obtain that $K_2 = [-1.7850, -1.1131]$. With the bounded compatible initial states, Fig. 5.5 shows the state curves of the singular swarm systems, where the state curves of the leaders and followers are denoted by the dotted lines and the solid lines, respectively, and the envelopes formed by the state curves of the leaders are marked by circles. One can see that the curves of followers stay in the envelope formed by those of the leaders; that is, the time-delayed singular swarm system achieves containment.

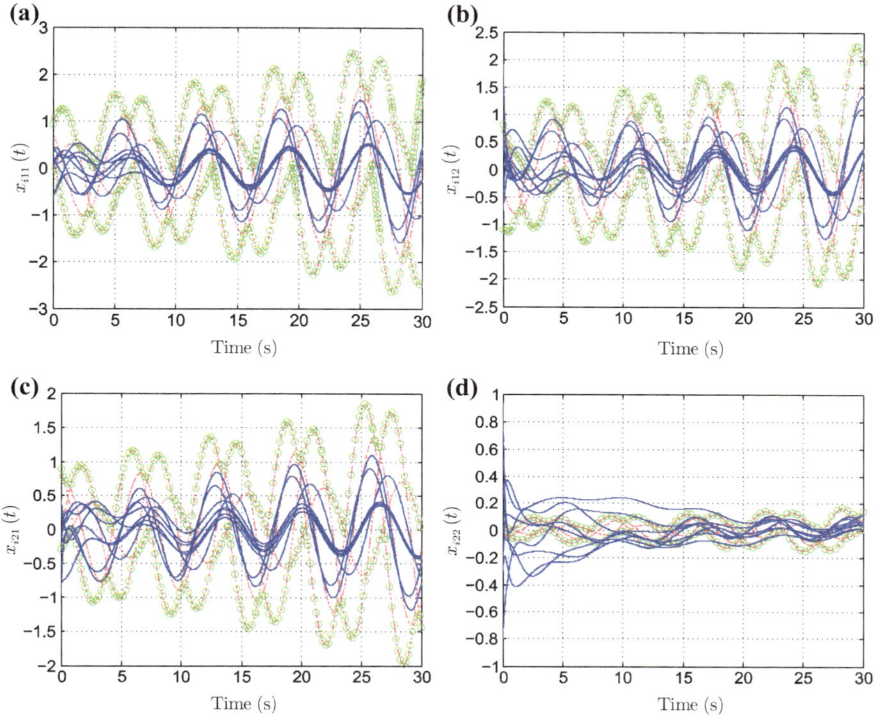

Fig. 5.5 Curves of $x_{i11}(t)$, $x_{i12}(t)$, $x_{i21}(t)$, and $x_{i22}(t)$ ($i = 1, 2, \ldots, 12$). **a** Curve of $x_{i11}(t)$. **b** Curve of $x_{i12}(t)$. **c** Curve of $x_{i21}(t)$. **d** Curve of $x_{i22}(t)$

5.4 Conclusions

In this chapter, output containment control problems for high-order LTI swarm systems were first studied using a dynamic output feedback approach. Necessary and sufficient conditions for swarm systems to achieve output containment were proposed and an approach to design the output containment protocol by solving an algebraic Reccati equation was given. Then state containment control problems for high-order singular LTI swarm systems with time delays were investigated. Sufficient conditions for swarm systems to achieve state containment were presented in terms of LMIs, which only include four LMI constraints independent of the number of agents. By using the method of changing variables, an approach was given to determine the gain matrices in the protocols. The results on output containment control can be applied to solve output consensus tracking and state containment control problems of swarm systems. The results on state containment control of singular swarm systems with time delays can be used to deal with the state containment control problems and state consensus tracking problems for normal and singular swarm systems with time delays. The materials of this chapter are mainly based on [20, 21]

It should be pointed out that, we also discussed the output containment control problems for high-order LTI swarm systems using static output feedback approaches, and proposed sufficient conditions for swarm systems to achieve output containment and approaches to design the protocol in [10]. Since this work can be regarded as a special case of the work in Sect. 6.3, it is omitted in this chapter.

References

1. Ji M, Ferrari-Trecate G, Egerstedt M et al (2008) Containment control in mobile networks. IEEE Trans Autom Control 53(8):1972–1975
2. Meng ZY, Ren W, You Z (2010) Distributed finite-time attitude containment control for multiple rigid bodies. Automatica 46(12):2092–2099
3. Notarstefano G, Egerstedt M, Haque M (2011) Containment in leader-follower networks with switching communication topologies. Automatica 47(5):1035–1040
4. Cao YC, Ren W, Egerstedt M (2012) Distributed containment control with multiple stationary or dynamic leaders in fixed and switching directed networks. Automatica 48(8):1586–1597
5. Cao YC, Stuart D, Ren W et al (2011) Distributed containment control for multiple autonomous vehicles with double-integrator dynamics: algorithms and experiments. IEEE Trans Control Syst Technol 19(4):929–938
6. Liu HY, Xie GM, Wang L (2012) Necessary and sufficient conditions for containment control of networked multi-agent systems. Automatica 48(7):1415–1422
7. Lou YC, Hong YG (2012) Target containment control of multi-agent systems with random switching interconnection topologies. Automatica 48(5):879–885
8. Liu HY, Xie GM, Wang L (2012) Containment of linear multi-agent systems under general interaction topologies. Syst Control Lett 61(4):528–534
9. Li ZK, Ren W, Liu XD et al (2013) Distributed containment control of multi-agent systems with general linear dynamics in the presence of multiple leaders. Int J Robust Nonlinear Control 23(5):534–547
10. Dong XW, Shi ZY, Lu G et al (2015) Output containment analysis and design for high-order linear time-invariant swarm systems. Int J Robust Nonlinear Control 25(6):900–913
11. Ni W, Cheng DZ (2010) Leader-following consensus of multi-agent systems under fixed and switching topologies. Syst Control Lett 59(3–4):209–217
12. Dai L (1989) Singular control systems. Springer, Berlin
13. Artstein Z (1982) Linear systems with delayed control: a reduction. IEEE Trans Autom Control 27(4):869–879
14. Gu K, Niculescu S (2003) Survey on recent results in the stability and control of time-delay systems. J Dyn Syst Meas Control 125(1):158–165
15. Richard J (2003) Time-delay systems: an overview of some recent advances and open problems. Automatica 39(10):1667–1694
16. Mondie S, Michiels W (2003) Finite spectrum assignment of unstable time-delay systems with a safe implementation. IEEE Trans Autom Control 48(12):2207–2212
17. Zhang XM, Wu M, She JH et al (2005) Delay-dependent stabilization of linear systems with time-varying state and input delays. Automatica 41(8):1405–1412
18. Wu L, Su X, Shi P et al (2011) Model approximation for discrete-time state-delay systems in the T-S fuzzy framework. IEEE Trans Fuzzy Syst 19(2):366–378
19. Wu L, Su X, Shi P et al (2011) A new approach to stability analysis and stabilization of discrete-time T-S fuzzy time-varying delay systems. IEEE Trans Syst Man Cybern Part B-Cybern 41(1):273–286

20. Dong XW, Meng FL, Shi ZY et al (2014) Output containment control for swarm systems with general linear dynamics: a dynamic output feedback approach. Syst Control Lett 71(1):31–37
21. Dong XW, Xi JX, Lu G et al (2014) Containment analysis and design for high-order linear time-invariant singular swarm systems with time delays. Int J Robust Nonlinear Control 24(7):1189–1204

Chapter 6
Formation-Containment Control of Swarm Systems

Abstract This chapter studies state and output formation-containment control problems for high-order linear time-invariant (LTI) swarm systems. First, protocols are presented for leaders and followers, respectively, to drive the states of leaders to realize the predefined time-varying state formation and propel the states of followers, to converge to the convex hull formed by states of leaders. State formation-containment problems of swarm systems are transformed into asymptotic stability problems, and an explicit expression of the formation reference function is derived. Sufficient conditions for swarm systems to achieve state formation-containment are proposed. Necessary and sufficient conditions for swarm systems to achieve state containment and time-varying state formation are presented respectively as special cases. An approach to determine the gain matrices in the state formation-containment protocols is given. Then, output formation-containment control problems for high-order LTI swarm systems are dealt with. Sufficient conditions for swarm systems to achieve output formation-containment are proposed. An approach to determine the gain matrices in the output formation-containment protocols for swarm systems to achieve output formation-containment is given. It is revealed that state formation-containment, state/output containment, state/output formation control, state/output consensus, and state/output consensus tracking problems can be unified in the framework of output formation-containment problems. Finally, numerical simulations are provided to demonstrate theoretical results.

6.1 Introduction

In Chap. 5, containment control problems for high-order swarm systems are studied. It should be pointed out that in most of the existing results on containment, such as those in [1–10], it is assumed that there exists no coordination among leaders. In some practical application, leaders need to coordinate with each other to accomplish certain complicated tasks, such as formation control. A swarm system is said to achieve state/output formation-containment if the states/outputs of leaders achieve desired formation, and at the same time the states/outputs of followers converge to the

X. Dong, *Formation and Containment Control for High-order Linear Swarm Systems*, Springer Theses, DOI 10.1007/978-3-662-47836-3_6

convex hull formed by the states/outputs of leaders. In [11], formation-containment concept was proposed for first-order swarm systems. Dimarogonas et al. [12] studied formation-containment control problems for nonholonomic robot swarm systems and presented sufficient conditions to achieve formation-containment. However, in [11, 12], the dynamics of each agent is of low-order. Formation-containment control problems for general high-order swarm systems have not been investigated before. Moreover, consensus control, consensus tracking control, formation control, and containment control can all be regarded as special cases of formation-containment control. The research on formation-containment control depends on the theory on formation control and containment control. Therefore, it is significant and also very challenging to study the state and output formation-containment control problems for high-order LTI swarm systems.

Based on the results obtained in Chaps. 4 and 5, this chapter studies state and output formation-containment control problems for general high-order LTI swarm systems. For state formation-containment control problems, formation-containment control protocols are constructed for leaders and followers, respectively, using the neighboring state information. Then, state formation-containment control problems are transformed into stability problems, and sufficient conditions for swarm systems to achieve state formation-containment are proposed. Necessary and sufficient conditions for high-order LTI swarm systems to achieve state containment and time-varying state formation are presented as special cases. It is verified that state consensus control (e.g., [13]), state consensus tracking control (e.g., [14]), state containment control (e.g., [15, 16]), and state containment control (e.g., [4, 5]) can all be treated as special cases of the state formation-containment control studied in this chapter. Finally, approaches to design the state formation-containment protocol are proposed. For output formation-containment control problems, based on static output feedback approach, output formation-containment control protocols are constructed by the outputs of neighboring agents. Based on output transformation and partial stability theory, output formation-containment problems are transformed into static output stabilization problems. Sufficient conditions for swarm systems to achieve output formation-containment are proposed, and approaches to determine the gain matrices in the output formation-containment protocol are given. The results on output formation-containment control can also be applied to deal with state/output consensus, state/output consensus tracking, state/output formation, state/output containment, and state formation-containment control problems of swarm systems.

The rest of this chapter is organized as follows: Sect. 6.2 studies state formation-containment control problems for high-order swarm systems. Section 6.3 investigates output formation-containment control problems for high-order swarm systems. Section 6.4 concludes the whole work of this chapter.

6.2 State Formation-Containment Control

State formation-containment control problems for high-order LTI swarm systems are studied in this section. Sufficient conditions for swarm systems to achieve state formation-containment and approaches to design the protocol are proposed. Necessary and sufficient conditions for swarm systems to achieve state containment and time-varying state formation are presented as special cases.

6.2.1 Problem Description

Consider a high-order LTI swarm system with N agents described by

$$\dot{x}_i(t) = Ax_i(t) + Bu_i(t)\ (i = 1, 2, \ldots, N), \tag{6.1}$$

where $x_i(t) \in \mathbb{R}^n$ are the states and $u_i(t) \in \mathbb{R}^m$ are the control inputs.

Definition 6.1 Agents in the swarm system are classified into leaders and followers according to the tasks they need to accomplish. An agent is called a *leader* if it coordinates with its neighbors to achieve the desired formation. An agent is called a *follower* if it has at least one neighbor in the swarm system, and it coordinates with its neighbors to converge to the convex hull formed by the leaders. The neighbors of a leaders are only leaders while the neighbors of a follower can be both leaders and followers.

Similar to Chap. 5, it is assumed that there are M $(M < N)$ followers and $N - M$ leaders in swarm system (6.1). Assumption 1 is still required in this chapter. Besides, it is assumed that the interaction topology among leaders has a spanning tree.

Definition 6.2 Swarm system (6.1) is said to *achieve state containment* if for any given bounded initial states, and any $k \in \mathscr{F}$, there exist nonnegative constant $\beta_{k,j}$ $(j \in \mathscr{L})$ satisfying $\sum_{j=M+1}^{N} \beta_{k,j} = 1$ such that

$$\lim_{t\to\infty} \left(x_k(t) - \sum_{j=M+1}^{N} \beta_{k,j} x_j(t) \right) = 0. \tag{6.2}$$

Define the desired time-varying state formation for leaders as $h_{\mathscr{L}}(t) = [h_{M+1}^T(t), h_{M+2}^T(t), \ldots, h_N^T(t)]^T$, where $h_i(t)$ $(i \in \mathscr{L})$ is piecewise continuously differentiable.

Definition 6.3 Swarm system (6.1) is said to *achieve state formation-containment* if for any given bounded initial states, any $i \in \mathscr{L}$ and $k \in \mathscr{F}$ there exist a vector-valued function $r(t) \in \mathbb{R}^n$ and nonnegative constants $\beta_{k,j}$ satisfying $\sum_{j=M+1}^{N} \beta_{k,j} = 1$ such that (4.2) and (6.2) hold simultaneously.

Remark 6.1 From Definitions 4.1, 4.3, 6.2 and 6.3, one sees that if $M = 0$, state formation-containment problems become the state formation problems. If $M = N-1$ or $M = 0$ and $h_{\mathscr{L}}(t) \equiv 0$, state formation-containment problems are just the state consensus tracking problems or state consensus problems, respectively, and the state formation reference function is just the state consensus function. If $h_{\mathscr{L}}(t) \equiv 0$ and leaders have no neighbors; that is, for all $i, j \in \mathscr{L}$, $w_{ij} = 0$, the state formation-containment problem becomes the state containment problem. Therefore, state formation problems, state consensus tracking problems, state consensus problems, and state containment problems all can be regarded as special cases of state formation-containment problems.

Consider the following state formation-containment protocols

$$u_i(t) = K_1 x_i(t) + K_2 \sum_{j \in N_i} w_{ij} \left(x_i(t) - x_j(t)\right), \; i \in \mathscr{F}, \tag{6.3}$$

$$u_i(t) = K_1 x_i(t) + K_3 \sum_{j \in N_i} w_{ij} \left((x_i(t) - h_i(t)) - (x_j(t) - h_j(t))\right), \; i \in \mathscr{L}, \tag{6.4}$$

where $K_i \in \mathbb{R}^{m \times n}$ $(i = 1, 2, 3)$ are constant gain matrices.

Remark 6.2 In protocols (6.3) and (6.4), the gain matrix K_1 will be used to assign the motion modes of the state formation reference. K_2 and K_3 will be used to drive the states of followers to converge to the convex hull formed by those of leaders and propel the states of leaders to achieve the desired state formation, respectively. Gain matrices K_i $(i = 1, 2, 3)$ can be designed by the following procedure. First, design K_1 to assign the motion modes of the state formation reference. Then design K_2 and K_3 to achieve the state formation-containment. It should be pointed out that K_1 is not necessary for swarm systems to achieve some state formation-containment.

In the current section, the following two problems for swarm system (6.1) with protocols (6.3) and (6.4) are mainly investigated: (i) under what conditions state formation-containment can be achieved; and (ii) how to determine the gain matrices in protocols (5) and (6) to achieve the state formation-containment.

6.2.2 Problem Transformation and Primary Results

Denote by $G_{\mathscr{L}}$ the interaction topology among leaders. Under Definition 6.1, the Laplacian matrix corresponding to G has the following form:

$$L = \begin{bmatrix} L_1 & L_2 \\ 0 & L_3 \end{bmatrix},$$

where $L_1 \in \mathbb{R}^{M\times M}$, $L_2 \in \mathbb{R}^{M\times(N-M)}$, and $L_3 \in \mathbb{R}^{(N-M)\times(N-M)}$ which is the Laplacian matrix corresponding to $G_{\mathscr{L}}$. Let $x_{\mathscr{F}}(t) = [x_1^T(t), x_2^T(t), \ldots, x_M^T(t)]^T$ and $x_{\mathscr{L}}(t) = [x_{M+1}^T(t), x_{M+2}^T(t), \ldots, x_N^T(t)]^T$. Under protocols (6.3) and (6.4), swarm system (6.1) can be written in a compact form as follows:

$$\dot{x}_{\mathscr{F}}(t) = (I_M \otimes (A + BK_1) + L_1 \otimes BK_2)\, x_{\mathscr{F}}(t) + (L_2 \otimes BK_2)x_{\mathscr{L}}(t), \quad (6.5)$$

$$\dot{x}_{\mathscr{L}}(t) = (I_{N-M} \otimes (A + BK_1) + L_3 \otimes BK_3)\, x_{\mathscr{L}}(t) - (L_3 \otimes BK_3)\, h_{\mathscr{L}}(t). \quad (6.6)$$

Let $\tilde{x}_i(t) = x_i(t) - h_i(t)$ $(i \in \mathscr{L})$ and $\tilde{x}_{\mathscr{L}}(t) = [\tilde{x}_{M+1}^T(t), \tilde{x}_{M+2}^T(t), \ldots, \tilde{x}_N^T(t)]^T$. Then system (6.6) can be rewritten as follows:

$$\begin{aligned}\dot{\tilde{x}}_E(t) = (I_{N-M} \otimes (A + BK_1) + L_3 \otimes BK_3)\, \tilde{x}_E(t) + (I_{N-M} \otimes (A + BK_1))\\ h_{\mathscr{L}}(t) - (I_{N-M} \otimes I)\dot{h}_{\mathscr{L}}(t). \end{aligned} \quad (6.7)$$

The following lemma holds directly from the definitions of state formation and state consensus.

Lemma 6.1 *Swarm system (6.6) achieves time-varying state formation $h_{\mathscr{L}}(t)$ if and only if swarm system (6.7) achieves state consensus.*

Let λ_i $(i \in \mathscr{L})$ denote the eigenvalue of matrix L_3, where $\lambda_{M+1} = 0$ with associated eigenvector $\bar{u}_{M+1} = \mathbf{1}$ and $0 < \mathrm{Re}(\lambda_{M+2}) \le \cdots \le \mathrm{Re}(\lambda_N)$. Let $U_{\mathscr{L}}^{-1} L_3 U_{\mathscr{L}} = J_{\mathscr{L}}$, where $U_{\mathscr{L}} = [\bar{u}_{M+1}, \bar{u}_{M+2}, \ldots, \bar{u}_N]$, $U_{\mathscr{L}}^{-1} = [\tilde{u}_{M+1}, \tilde{u}_{M+2}, \ldots, \tilde{u}_N]^H$ and $J_{\mathscr{L}}$ is the Jordan canonical form of L_3. From Lemma 2.1 and the structure of $U_{\mathscr{L}}$, one can set $J_{\mathscr{L}} = \mathrm{diag}\{0, \bar{J}_{\mathscr{L}}\}$ where $\bar{J}_{\mathscr{L}}$ consists of Jordan blocks corresponding to λ_i $(i = M+2, M+3, \ldots, N)$. Let $\tilde{U}_{\mathscr{L}} = [\tilde{u}_{M+2}, \tilde{u}_{M+3}, \ldots, \tilde{u}_N]^H$, $\theta_{\mathscr{L}}(t) = (\tilde{u}_{M+1} \otimes I)\tilde{x}_{\mathscr{L}}(t)$ and $\varsigma(t) = (\tilde{U}_{\mathscr{L}} \otimes I)\tilde{x}_{\mathscr{L}}(t)$, then system (6.7) can be transformed into

$$\dot{\theta}_{\mathscr{L}}(t) = (A + BK_1)\, \theta_{\mathscr{L}}(t) + (\tilde{u}_{M+1} \otimes (A + BK_1))\, h_{\mathscr{L}}(t) - (\tilde{u}_{M+1} \otimes I)\dot{h}_{\mathscr{L}}(t), \quad (6.8)$$

$$\begin{aligned}\dot{\varsigma}(t) = \left(I_{N-M-1} \otimes (A + BK_1) + \bar{J}_{\mathscr{L}} \otimes BK_3\right) \varsigma(t) + (\tilde{U}_{\mathscr{L}} \otimes (A + BK_1))h_{\mathscr{L}}(t)\\ -(\tilde{U}_{\mathscr{L}} \otimes I)\dot{h}_{\mathscr{L}}(t). \end{aligned} \quad (6.9)$$

The following lemma presents a necessary and sufficient condition for swarm system (6.6) to achieve time-varying state formation $h_{\mathscr{L}}(t)$.

Lemma 6.2 *Swarm system (6.6) achieves time-varying state formation $h_{\mathscr{L}}(t)$ if and only if $\lim_{t\to\infty}\varsigma(t) = 0$.*

Proof Let

$$\tilde{x}_{\mathscr{L}C}(t) = \mathbf{1} \otimes \theta_{\mathscr{L}}(t), \quad (6.10)$$

$$\tilde{x}_{\mathscr{L}\bar{C}}(t) = \tilde{x}_{\mathscr{L}}(t) - \tilde{x}_{\mathscr{L}C}(t). \tag{6.11}$$

Let $\bar{e}_i \in \mathbb{R}^{N-M}$ $(i \in \mathscr{L})$ be a vector with 1 as its $i-M$th component and 0 elsewhere. Since $[\theta_{\mathscr{L}}^H(t), 0]^H = \bar{e}_{M+1} \otimes \theta_{\mathscr{L}}(t)$, one gets

$$\tilde{x}_{\mathscr{L}C}(t) = (U_{\mathscr{L}}\bar{e}_{M+1}) \otimes \theta_{\mathscr{L}}(t) = (U_{\mathscr{L}} \otimes I)[\theta_{\mathscr{L}}^H(t), 0]^H. \tag{6.12}$$

Because $[\theta_{\mathscr{L}}^H(t), \varsigma^H(t)]^H = (U_{\mathscr{L}}^{-1} \otimes I)\tilde{x}_{\mathscr{L}}(t)$, it can be obtained that

$$\tilde{x}_{\mathscr{L}\bar{C}}(t) = (U_{\mathscr{L}} \otimes I)[0, \varsigma^H(t)]^H. \tag{6.13}$$

Due to the fact that $U_{\mathscr{L}}^{-1} \otimes I$ is nonsingular, from (6.12) and (6.13), one knows that $\tilde{x}_{\mathscr{L}C}(t)$ and $\tilde{x}_{\mathscr{L}\bar{C}}(t)$ are linearly independent. Therefore, from (6.10) and (6.11), it holds that the subsystems with states $\tilde{x}_{\mathscr{L}C}(t)$ and $\tilde{x}_{\mathscr{L}\bar{C}}(t)$ describe the consensus dynamics and disagreement dynamics of swarm system (6.7), respectively. From Lemma 6.1, one sees that swarm system (6.6) achieves time-varying formation $h_{\mathscr{L}}(t)$ if and only if $\lim_{t\to\infty}\tilde{x}_{\mathscr{L}\bar{C}}(t) = 0$, which also means that $\lim_{t\to\infty}\varsigma(t) = 0$. This completes the proof.

Let

$$\xi_{\mathscr{F}i}(t) = \sum_{j \in N_i} w_{ij} \left(x_i(t) - x_j(t)\right) \; (i \in \mathscr{F}),$$

$\xi_{\mathscr{F}}(t) = [\xi_{\mathscr{F}1}^T(t), \xi_{\mathscr{F}2}^T(t), \ldots, \xi_{\mathscr{F}M}^T(t)]^T$. Then

$$\xi_{\mathscr{F}}(t) = (L_2 \otimes I)\, x_{\mathscr{L}}(t) + (L_1 \otimes I)\, x_{\mathscr{F}}(t). \tag{6.14}$$

If $\lim_{t\to\infty}\xi_{\mathscr{F}}(t) = 0$, one can obtain

$$\lim_{t\to\infty} \left(x_{\mathscr{F}}(t) - \left(-L_1^{-1}L_2 \otimes I\right) x_{\mathscr{L}}(t)\right) = 0. \tag{6.15}$$

From (6.15) and Lemmas 5.1 and 6.2, the following lemma holds directly.

Lemma 6.3 *Swarm systems (6.1) under protocols (6.3) and (6.4) achieves formation-containment if for any given bounded initial states*

$$\begin{cases} \lim\limits_{t\to\infty} \varsigma(t) = 0, \\ \lim\limits_{t\to\infty} \xi_{\mathscr{F}}(t) = 0. \end{cases}$$

Based on the above analysis, an explicit expression of the state formation reference function can be presented.

Lemma 6.4 *If swarm systems (6.1) under protocols (6.3) and (6.4) achieve state formation-containment, then the state formation reference function $r(t)$ satisfies*

$$\lim_{t\to\infty}(r(t) - r_0(t) - r_h(t)) = 0,$$

with $r_0(t) = e^{(A+BK_1)t}\left(\tilde{u}_{M+1}^H \otimes I\right) x_{\mathscr{L}}(0)$ *and* $r_h(t) = -(\tilde{u}_{M+1}^H \otimes I)h_{\mathscr{L}}(t)$.

Proof If swarm systems (6.1) under protocols (6.3) and (6.4) achieve state formation-containment, then $\lim_{t\to\infty}\varsigma(t) = 0$ and

$$\lim_{t\to\infty}(\tilde{x}_{\mathscr{F}}(t) - \mathbf{1} \otimes \theta_{\mathscr{L}}(t)) = 0. \tag{6.16}$$

Therefore, subsystem (6.8) determines the formation reference function. It can be shown that

$$\begin{aligned}
&\int_0^t e^{(A+BK_1)(t-\tau)}(\tilde{u}_{M+1}^H \otimes I)\dot{h}_{\mathscr{L}}(\tau)d\tau \\
&\quad = (\tilde{u}_{M+1}^H \otimes I)h_{\mathscr{L}}(t) - e^{(A+BK_1)t}(\tilde{u}_{M+1}^H \otimes I)h_{\mathscr{L}}(0) \\
&\quad + \int_0^t e^{(A+BK_1)(t-\tau)}\left(\tilde{u}_{M+1}^H \otimes (A+BK_1)\right)h_{\mathscr{L}}(\tau)d\tau,
\end{aligned} \tag{6.17}$$

and

$$\theta_{\mathscr{L}}(0) = \left(\tilde{u}_{M+1}^H \otimes I\right)(x_{\mathscr{L}}(0) - h_{\mathscr{L}}(0)). \tag{6.18}$$

From (6.8) and (6.16)–(6.18), the conclusion of Lemma 6.4 can be obtained.

Remark 6.3 In Lemma 6.4, $r_0(t)$ is said to be the formation reference function of the swarm system with state formation $h_{\mathscr{L}}(t) \equiv 0$. $r_h(t)$ describes the impact of the state formation $h_{\mathscr{L}}(t)$. From Lemma 6.4, one sees that the gain matrix K_1 can be used to assign motion modes of the formation reference by specifying the eigenvalues of $(A + BK_1)$ at the desired locations in the complex plane.

6.2.3 State Formation-Containment Analysis and Protocol Design

Using the same $U_{\mathscr{F}}$ and $\bar{\Lambda}_{\mathscr{F}}$ as the ones in Sect. 5.2, one can obtain the following sufficient conditions for swarm systems to achieve state formation-containment.

Theorem 6.1 *For any given bounded initial states, swarm system (6.1) under protocols (6.3) and (6.4) achieves state formation-containment if the following conditions hold simultaneously*

(i) For all $i \in \mathscr{L}$

$$\lim_{t\to\infty}\left((A+BK_1)\left(h_i(t) - h_j(t)\right) - \left(\dot{h}_i(t) - \dot{h}_j(t)\right)\right) = 0, \ j \in N_i; \tag{6.19}$$

(ii) For all $i \in \mathscr{F}$, $A + BK_1 + \lambda_i BK_2$ are Hurwitz;
(iii) For all $i \in \{M+2, M+3, \ldots, N\}$, $A + BK_1 + \lambda_i BK_3$ are Hurwitz.

Proof If condition (i) holds, one can obtain that

$$\lim_{t\to\infty} \left((L_3 \otimes (A+BK_1))\, h_{\mathscr{L}}(t) - (L_3 \otimes I)\, \dot{h}_{\mathscr{L}}(t)\right) = 0. \tag{6.20}$$

Substituting $L_3 = U_{\mathscr{L}} J_{\mathscr{L}} U_{\mathscr{L}}^{-1}$ into (6.20) and pre-multiplying the left and right sides of (6.20) by $U_{\mathscr{L}}^{-1} \otimes I$ lead to

$$\lim_{t\to\infty} \left(\left(\bar{J}_{\mathscr{L}} \tilde{U}_{\mathscr{L}} \otimes (A + BK_1)\right) h_{\mathscr{L}}(t) - \left(\bar{J}_{\mathscr{L}} \tilde{U}_{\mathscr{L}} \otimes I\right) \dot{h}_{\mathscr{L}}(t)\right) = 0. \tag{6.21}$$

Since $G_{\mathscr{L}}$ has a spanning tree, by Lemma 2.1 and the structure of $J_{\mathscr{L}}$, one knows that $\bar{J}_{\mathscr{L}}$ is nonsingular. Pre-multiplying the left and right sides of (6.21) by $\bar{J}_{\mathscr{L}}^{-1} \otimes I$, one has

$$\lim_{t\to\infty} \left(\left(\tilde{U}_{\mathscr{L}} \otimes (A + BK_1)\right) h_{\mathscr{L}}(t) - \left(\tilde{U}_{\mathscr{L}} \otimes I\right) \dot{h}_{\mathscr{L}}(t)\right) = 0. \tag{6.22}$$

Consider the following $N - M - 1$ subsystems

$$\dot{\tilde{\theta}}_{\mathscr{L}i}(t) = (A + BK_1 + \lambda_i BK_3)\, \tilde{\theta}_{\mathscr{L}i}(t) \ (i = M+2, M+3, \ldots, N). \tag{6.23}$$

If condition (iii) holds, then the $N - M - 1$ subsystems described by (6.23) are asymptotically stable. In this case, it holds that the system described by

$$\dot{\tilde{\varsigma}}(t) = \left(I_{N-M-1} \otimes (A + BK_1) + \bar{J}_{\mathscr{L}} \otimes BK_3\right) \tilde{\varsigma}(t) \tag{6.24}$$

is asymptotically stable. From (6.9), (6.22) and (6.24), one knows that

$$\lim_{t\to\infty} \varsigma(t) = 0. \tag{6.25}$$

By (6.25) and Lemma 6.2, swarm system (6.6) achieves time-varying state formation $h_{\mathscr{L}}(t)$.

When the leaders achieve time-varying formation $h_{\mathscr{L}}(t)$, then

$$\lim_{t\to\infty} ((L_2 L_3) \otimes (BK_3))\, (x_{\mathscr{L}}(t) - h_{\mathscr{L}}(t)) = \lim_{t\to\infty} ((L_2 L_3) \otimes (BK_3))\, (\mathbf{1} \otimes r(t)). \tag{6.26}$$

Since $L_3 \mathbf{1} = 0$, one has

$$\lim_{t\to\infty} ((L_2 L_3) \otimes (BK_3))\, (x_{\mathscr{L}}(t) - h_{\mathscr{L}}(t)) = 0. \tag{6.27}$$

If condition (ii) holds, then it can be shown that for all $i \in \mathscr{F}$, the following M subsystems

$$\dot{\tilde{\xi}}_{\mathscr{F}i}(t) = (A + BK_1 + \lambda_i BK_2)\,\tilde{\xi}_{\mathscr{F}i}(t) \tag{6.28}$$

are asymptotically stable; that is, the following system

$$\dot{\tilde{\xi}}_{\mathscr{F}}(t) = (I_M \otimes (A + BK_1) + (L_1 \otimes BK_2))\,\tilde{\xi}_{\mathscr{F}}(t) \tag{6.29}$$

is asymptotically stable. From (6.5), (6.6), and (6.14), it can be derived that

$$\begin{aligned}\dot{\xi}_{\mathscr{F}}(t) = {} & (I_M \otimes (A + BK_1) + (L_1 \otimes BK_2))\,\xi_{\mathscr{F}}(t) + ((L_2 L_3) \otimes (BK_3))\\ & (x_{\mathscr{L}}(t) - h_{\mathscr{L}}(t))\,.\end{aligned} \tag{6.30}$$

From (6.27), (6.29) and (6.30), it holds that

$$\lim_{t\to\infty} \xi_{\mathscr{F}}(t) = 0. \tag{6.31}$$

From (6.25), (6.31), and Lemma 6.3, one sees that swarm system (6.1) under protocols (6.3) and (6.4) achieves state formation-containment. The proof of Theorem 6.1 is completed.

Theorem 6.1 presents sufficient conditions to achieve state formation-containment. In the case where leaders have no neighbors, that is, for all $i, j \in \mathscr{L}$, $w_{ij} = 0$, state formation-containment problems become state containment problems, and the following Corollary can be obtained.

Corollary 6.1 *In the case where leaders have no neighbors, for any given bounded initial states, swarm system (6.1) under protocols (6.3) and (6.4) achieves state containment if and only if condition (ii) in Theorem 6.1 holds.*

Proof Sufficiency can be obtained directly from the proof of Theorem 6.1. The necessary is proven by contradiction. Suppose that swarm system (6.1) under protocols (6.3) and (6.4) achieves state containment for any given bounded initial states and condition (ii) in Theorem 6.1 does not hold. Since leaders have no neighbors, then $L_3 = 0$. Let $x(t) = [x_{\mathscr{F}}^T(t), x_{\mathscr{L}}^T(t)]^T$. Systems (6.5) and (6.6) can be rewritten in a compact form as follows:

$$\dot{x}(t) = \begin{bmatrix} \tilde{\Psi}_1 & \tilde{\Psi}_2 \\ 0 & \tilde{\Psi}_3 \end{bmatrix} x(t), \tag{6.32}$$

where

$$\tilde{\Psi}_1 = I_M \otimes (A + BK_1) + L_1 \otimes BK_2,\ \tilde{\Psi}_2 = L_2 \otimes BK_2,\ \tilde{\Psi}_3 = I_{N-M} \otimes (A + BK_1).$$

The solution to equation (6.32) can be written as

$$x(t) = \begin{bmatrix} e^{\tilde{\Psi}_1 t} & \mathcal{L}^{-1}\left(\left(sI - \tilde{\Psi}_1\right)^{-1} \tilde{\Psi}_2 \left(sI - \tilde{\Psi}_3\right)^{-1}\right) \\ 0 & e^{\tilde{\Psi}_3 t} \end{bmatrix} \begin{bmatrix} x_{\mathscr{F}}(0) \\ x_{\mathscr{L}}(0) \end{bmatrix}. \tag{6.33}$$

It can be shown that

$$\begin{aligned} \left(sI - \tilde{\Psi}_1\right)^{-1} \tilde{\Psi}_2 \left(sI - \tilde{\Psi}_3\right)^{-1} = & \left(sI - \tilde{\Psi}_1\right)^{-1} \left(L_1^{-1} L_2 \otimes I\right) \\ & + \left(-L_1^{-1} L_2 \otimes I\right) \left(sI - \tilde{\Psi}_3\right)^{-1}. \end{aligned}$$

So one has

$$x(t) = \begin{bmatrix} e^{\tilde{\Psi}_1 t} & e^{\tilde{\Psi}_1 t}\left(L_1^{-1} L_2 \otimes I\right) + \left(-L_1^{-1} L_2 \otimes I\right) e^{\tilde{\Psi}_3 t} \\ 0 & e^{\tilde{\Psi}_3 t} \end{bmatrix} \begin{bmatrix} x_{\mathscr{F}}(0) \\ x_{\mathscr{L}}(0) \end{bmatrix}. \tag{6.34}$$

From (6.34), one obtains

$$x_{\mathscr{F}}(t) = e^{\tilde{\Psi}_1 t} x_{\mathscr{F}}(0) + e^{\tilde{\Psi}_1 t}\left(L_1^{-1} L_2 \otimes I\right) x_{\mathscr{L}}(0) + \left(-L_1^{-1} L_2 \otimes I\right) e^{\tilde{\Psi}_3 t} x_{\mathscr{L}}(0). \tag{6.35}$$

By the structure of $U_{\mathscr{F}}$ and $\bar{\Lambda}_{\mathscr{F}}$, if condition (ii) in Theorem 6.1 does not hold, then $\tilde{\Psi}_1$ is not Hurwitz. Choose $x_{\mathscr{L}}(0) = 0$ and $x_{\mathscr{F}}(0) \neq 0$. Then $x_{\mathscr{L}}(t) = 0$ but $\lim_{t \to \infty} x_{\mathscr{F}}(t) \neq 0$, which means that swarm system (6.1) under protocols (6.3) and (6.4) does not achieve state containment. This results a contradiction. The proof of Corollary 6.1 is completed.

Remark 6.4 Corollary 6.1 presents necessary and sufficient conditions for high-order LTI swarm systems with multiple leaders to achieve containment. To the best of our knowledge, for high-order swarm systems with multiple leaders only sufficient conditions have been obtained before. Necessary and sufficient conditions for first-order and second-order swarm systems to achieve containment shown in [4, 5] can be regarded as special cases of Corollary 6.1.

If $M = 0$, state formation-containment problems become state formation problems, and the following corollary can be obtained.

Corollary 6.2 *In the case where $M = 0$, for any given bounded initial states, swarm system (6.1) under the protocol (6.4) achieves time-varying state formation $h_{\mathscr{L}}(t)$ if and only if conditions (i) and (iii) in Theorem 6.1 hold.*

Remark 6.5 It can be verified that by choosing appropriate A, B, $h_{\mathscr{L}}(t)$, and K_1, Corollary 6.2 is equivalent to the conclusions in [15]. If $K_1 = 0$ and $\dot{h}_{\mathscr{L}}(t) \equiv 0$, Corollary 6.2 becomes the Theorem 1 in [16].

Remark 6.6 If $M = 0$ and $h_{\mathscr{L}}(t) \equiv 0$, the state formation-containment problem in this section becomes the state consensus problem discussed in [13] and Theorem 1 in [13] can be treated as a special case of Corollary 6.2.

It should be mentioned that Theorem 6.1 includes $N - 1$ Hurwitz constraints. Therefore, when a swarm system consists of numerous agents, especially numerous followers, it may be difficult to check these Hurwitz constraints. The following theorem presents a method to improve the calculation efficiency by encapsulating all complex eigenvalues of L_1 and L_3 into convex sets, respectively. Let $\tilde{\lambda}_{1,2} = \mathrm{Re}(\lambda_{M+2}) \pm j\mu_{\mathscr{L}}$, $\tilde{\lambda}_{3,4} = \mathrm{Re}(\lambda_N) \pm j\mu_{\mathscr{L}}$, $\bar{\lambda}_{1,2} = \mathrm{Re}(\lambda_1) \pm j\mu_{\mathscr{F}}$ and $\bar{\lambda}_{3,4} = \mathrm{Re}(\lambda_M) \pm j\mu_{\mathscr{F}}$, where $\mu_{\mathscr{L}} = \max\{\mathrm{Im}(\lambda_i), i = M+2, M+3, \ldots, N\}$ and $\mu_{\mathscr{F}} = \max\{\mathrm{Im}(\lambda_i), i \in \mathscr{F}\}$.

Theorem 6.2 *For any given bounded initial states, swarm system (6.1) under protocols (6.3) and (6.4) achieves the state formation-containment if the following conditions hold simultaneously*
(i) For all $i \in \mathscr{L}$

$$\lim_{t\to\infty} \left((A + BK_1)\left(h_i(t) - h_j(t)\right) - \left(\dot{h}_i(t) - \dot{h}_j(t)\right)\right) = 0, \ \ j \in N_i;$$

(ii) There exists a matrix $\Lambda_{\bar{R}_{\mathscr{F}}} = \Lambda_{\bar{R}_{\mathscr{F}}}^T > 0$ *such that*

$$\left(\Lambda_{A+BK_1} + \Psi_{\bar{\lambda}_i}\Lambda_{BK_2}\right)^T \Lambda_{\bar{R}_{\mathscr{F}}} + \Lambda_{\bar{R}_{\mathscr{F}}}^T \left(\Lambda_{A+BK_1} + \Psi_{\bar{\lambda}_i}\Lambda_{BK_2}\right) < 0 \ (i = 1, 3);$$

(iii) There exists a matrix $\Lambda_{\tilde{R}_{\mathscr{L}}} = \Lambda_{\tilde{R}_{\mathscr{L}}}^T > 0$ *such that*

$$\left(\Lambda_{A+BK_1} + \Psi_{\tilde{\lambda}_i}\Lambda_{BK_3}\right)^T \Lambda_{\tilde{R}_{\mathscr{L}}} + \Lambda_{\tilde{R}_{\mathscr{L}}}^T \left(\Lambda_{A+BK_1} + \Psi_{\tilde{\lambda}_i}\Lambda_{BK_3}\right) < 0 \ (i = 1, 3).$$

Proof Let $\Gamma_{\mathscr{F}i} = \left(\Lambda_{A+BK_1} + \Psi_{\lambda_i}\Lambda_{BK_2}\right)^T \Lambda_{\bar{R}_{\mathscr{F}}} + \Lambda_{\bar{R}_{\mathscr{F}}}^T \left(\Lambda_{A+BK_1} + \Psi_{\lambda_i}\Lambda_{BK_2}\right)$ $(i \in \mathscr{F})$, then one can obtain

$$\Gamma_{\mathscr{F}i} = \bar{\Omega}_{\mathscr{F}0} + \mathrm{Re}(\lambda_i)\bar{\Omega}_{\mathscr{F}1} + \mathrm{Im}(\lambda_i)\bar{\Omega}_{\mathscr{F}2},$$

where

$$\bar{\Omega}_{\mathscr{F}0} = \begin{bmatrix} (A + BK_1)^T \bar{R}_{\mathscr{F}} + \bar{R}_{\mathscr{F}}^T (A + BK_1) & 0 \\ 0 & (A + BK_1)^T \bar{R}_{\mathscr{F}} + \bar{R}_{\mathscr{F}}^T (A + BK_1) \end{bmatrix},$$

$$\bar{\Omega}_{\mathscr{F}1} = \begin{bmatrix} (BK_2)^T \bar{R}_{\mathscr{F}} + \bar{R}_{\mathscr{F}}^T BK_2 & 0 \\ 0 & (BK_2)^T \bar{R}_{\mathscr{F}} + \bar{R}_{\mathscr{F}}^T BK_2 \end{bmatrix},$$

$$\bar{\Omega}_{\mathscr{F}2} = \begin{bmatrix} 0 & (BK_2)^T \bar{R}_{\mathscr{F}} - \bar{R}_{\mathscr{F}}^T BK_2 \\ -(BK_2)^T \bar{R}_{\mathscr{F}} + \bar{R}_{\mathscr{F}}^T BK_2 & 0 \end{bmatrix}.$$

Similarly, let $\bar{\Gamma}_{\mathscr{F}i} = \left(\Lambda_{A+BK_1} + \Psi_{\bar{\lambda}_i}\Lambda_{BK_2}\right)^T \Lambda_{\bar{R}_{\mathscr{F}}} + \Lambda_{\bar{R}_{\mathscr{F}}}^T \left(\Lambda_{A+BK_1} + \Psi_{\bar{\lambda}_i}\Lambda_{BK_2}\right)$ $(i = 1, 2, 3, 4)$, then it follows that $\bar{\Gamma}_{\mathscr{F}i} = \bar{\Omega}_{\mathscr{F}0} + \mathrm{Re}(\bar{\lambda}_i)\bar{\Omega}_{\mathscr{F}1} + \mathrm{Im}(\bar{\lambda}_i)\bar{\Omega}_{\mathscr{F}2}$. Define

$$\bar{T}_{\mathscr{F}} = \begin{bmatrix} 0 & I \\ I & 0 \end{bmatrix}.$$

Owing to that $\mathrm{Im}\left(\bar{\lambda}_1\right) = -\mathrm{Im}\left(\bar{\lambda}_2\right)$ and $\mathrm{Im}\left(\bar{\lambda}_2\right) = -\mathrm{Im}\left(\bar{\lambda}_4\right)$, one has $\Pi_1 = \bar{T}_{\mathscr{F}}\Pi_2\bar{T}_{\mathscr{F}}^{-1}$ and $\Pi_3 = \bar{T}_{\mathscr{F}}\Pi_4\bar{T}_{\mathscr{F}}^{-1}$, which means that if $\bar{\Gamma}_{\mathscr{F}1} < 0$ and $\bar{\Gamma}_{\mathscr{F}3} < 0$, then $\bar{\Gamma}_{\mathscr{F}2} < 0$ and $\bar{\Gamma}_{\mathscr{F}4} < 0$ respectively. Therefore, by Lemma 4.3, one knows that if condition (ii) in Theorem 6.2 holds, then $\Gamma_{\mathscr{F}i} < 0$ $(i \in \mathscr{F})$.

By the decomposition of real and imaginary parts, it can verified that the asymptotic stabilities of subsystems (6.28) are equivalent to those of the following subsystems

$$\dot{\bar{\zeta}}_i(t) = \left(\Lambda_{A+BK_1} + \Psi_{\lambda_i}\Lambda_{BK_2}\right)\bar{\zeta}_i(t) \ (i \in \mathscr{F}). \tag{6.36}$$

Consider the stabilities of subsystems (6.36) . Choose a Lyapunov functional candidate as follows:

$$\bar{V}_i(t) = \bar{\zeta}_i^T(t)\Lambda_{\bar{R}_{\mathscr{F}}}\bar{\zeta}_i(t),$$

where $\Lambda_{\bar{R}_{\mathscr{F}}} = \Lambda_{\bar{R}_{\mathscr{F}}}^T > 0$ and $i \in \mathscr{F}$. Taking the time derivative of $\bar{V}_i(t)$ along the trajectory of (6.36), one has

$$\begin{aligned}\dot{\bar{V}}_i(t) = {} & \bar{\zeta}_i^T(t)\left(\Lambda_{A+BK_1} + \Psi_{\lambda_i}\Lambda_{BK_2}\right)^T \Lambda_{\bar{R}_{\mathscr{F}}}\bar{\zeta}_i(t) \\ & + \bar{\zeta}_i^T(t)\Lambda_{\bar{R}_{\mathscr{F}}}\left(\Lambda_{A+BK_1} + \Psi_{\lambda_i}\Lambda_{BK_2}\right)\bar{\zeta}_i(t).\end{aligned}$$

Since $\Gamma_{\mathscr{F}i} < 0$ $(i \in \mathscr{F})$, it can be verified that $\dot{\bar{V}}_i(t) < 0$, which means that condition (ii) in Theorem 6.1 holds. By a similar analysis as for condition (ii) in Theorem 6.2, it can be shown that condition (iii) in Theorem 6.2 guarantees condition (iii) in Theorem 6.1. Therefore, from Theorem 6.1, one knows that swarm system (6.1) under protocols (6.3) and (6.4) achieves formation-containment. This completes the proof.

Remark 6.7 One of the main properties of swarm systems is of large scale, so the solvability of the presented method is important. By the conclusions of Theorem 6.2, only four LMI constraints independent of the number of agents are required to be checked. Therefore, for very large N, especially for very large M, the calculation complexity can be decreased greatly.

Theorem 6.3 *If condition (i) in Theorem 6.1 holds and* (A, B) *is stabilizable, then for any given bounded initial states, swarm system (6.1) achieves state formation-containment by protocols (6.3) and (6.4) with*

$$K_2 = -[\mathrm{Re}(\lambda_1)]^{-1}\tilde{R}_o^{-1}B^T\tilde{P}_o,\ K_3 = -[\mathrm{Re}(\lambda_{M+2})]^{-1}\tilde{R}_o^{-1}B^T\tilde{P}_o,$$

where $\tilde{P}_o$ is the positive definite solution to the algebraic Riccati equation

$$\tilde{P}_o(A + BK_1) + (A + BK_1)^T\tilde{P}_o - \tilde{P}_oB\tilde{R}_o^{-1}B^T\tilde{P}_o + \tilde{Q}_o = 0, \tag{6.37}$$

for $\tilde{R}_o^T = \tilde{R}_o > 0$ and $\tilde{Q}_o = \tilde{D}_o^T\tilde{D}_o \geq 0$ with $(A + BK_1, \tilde{D}_o)$ detectable.

Proof If (A, B) is stabilizable, then $(A + BK_1, B)$ is stabilizable. Thus, for any given $\tilde{R}_o^T = \tilde{R}_o > 0$ and $\tilde{Q}_o = \tilde{D}_o^T\tilde{D}_o \geq 0$ with $(A + BK_1, \tilde{D}_o)$ detectable, algebraic Riccati equation (6.37) has a unique solution $\tilde{P}_o^T = \tilde{P}_o > 0$. Consider the $N - M - 1$ subsystems described by (6.23) and construct the following Lyapunov function candidates

$$\tilde{V}_i(t) = \tilde{\theta}_{\mathscr{L}i}^H(t)\tilde{P}_o\tilde{\theta}_{\mathscr{L}i}(t)\ (i = M + 2, M + 3, \ldots, N). \tag{6.38}$$

Let $K_3 = -[\mathrm{Re}(\lambda_{M+2})]^{-1}\tilde{R}_o^{-1}B^T\tilde{P}_o$. Taking the derivative of $\tilde{V}_i(t)$ with respect to t along the solution to subsystems (6.23), one has

$$\begin{aligned}\dot{\tilde{V}}_i(t) = &-\tilde{\theta}_{\mathscr{L}i}^H(t)\tilde{Q}_o\tilde{\theta}_{\mathscr{L}i}(t) + \left(1 - 2\mathrm{Re}(\lambda_i)[\mathrm{Re}(\lambda_{M+2})]^{-1}\right)\\ &\tilde{\theta}_{\mathscr{L}i}^H(t)\tilde{P}_oB\tilde{R}_o^{-1}B^T\tilde{P}_o\tilde{\theta}_{\mathscr{L}i}(t) \leq 0.\end{aligned} \tag{6.39}$$

Since $(A + BK_1, \tilde{D}_o)$ is detectable and $\tilde{R}_o^T = \tilde{R}_o > 0$, one can obtain that the $N - M - 1$ subsystems described by (6.23) are asymptotically stable.

Consider the M subsystems described by (6.28). By a similar analysis as for subsystems described by (6.23), it can be shown that $K_2 = -[\mathrm{Re}(\lambda_1)]^{-1}\tilde{R}_o^{-1}B^T\tilde{P}_o$ can make the M subsystems described by (6.28) asymptotically stable. From the proof of Theorem 6.1, one can see that swarm system (6.1) under protocols (6.3) and (6.4) achieves state formation-containment. This completes the proof.

Remark 6.8 If $M = N-1$, Corollary 6.1 presents necessary and sufficient conditions for swarm system (6.1) under protocols (6.3) and (6.4) to achieve consensus tracking. More specifically, if $M = N - 1$, $K_1 = 0$, $u_N(t) \equiv 0$ and the interaction topologies are undirected, then the formation-containment problems in the current paper become the consensus tracking problems investigated in [14]. Moreover, if $\tilde{R}_o^{-1} = 2\lambda_1$ and $\tilde{D}_o = \sqrt{\lambda_1}I$, Theorem 1 in [14] becomes a special case of Theorem 6.3 in this paper.

6.2.4 Numerical Simulations

Consider a swarm system with fourteen agents. The dynamics of each agent is described by (6.1) with

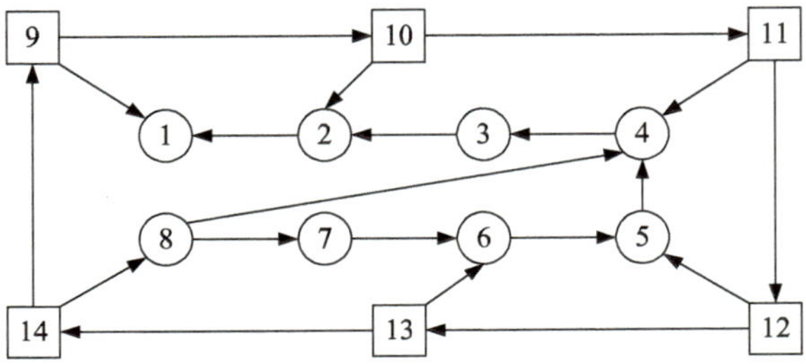

Fig. 6.1 Directed interaction topology G

$$A = \begin{bmatrix} -2.1 & 1 & -2.1 \\ 0 & 0 & 1 \\ -2 & 4 & 5 \end{bmatrix}, \ B = \begin{bmatrix} 0 \\ 0 \\ 1 \end{bmatrix}.$$

Assume that there are six leaders and eight followers in the swarm system. The interaction topology of the swarm system is shown in Fig. 6.1.

The state formation-containment problem for the swarm system is described as follows. First, the six leaders are required to preserve a periodic time-varying parallel

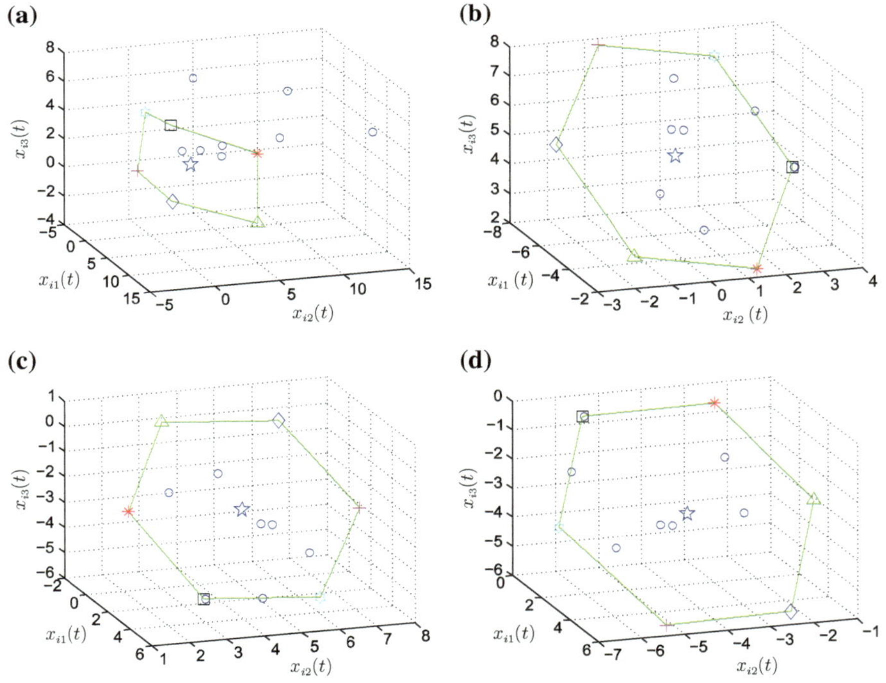

Fig. 6.2 Trajectory snapshots of fourteen agents and $r(t)$ at different time. **a** $t = 0$ s, **b** $t = 196$ s, **c** $t = 198$ s, **d** $t = 200$ s

hexagon formation and keep rotation around the predefined time-varying formation reference. The time-varying state formation is defined as follows:

$$h_i(t) = \begin{bmatrix} 3\sin\left(t + \frac{(i-9)\pi}{3}\right) \\ 3\cos\left(t + \frac{(i-9)\pi}{3}\right) \\ -3\sin\left(t + \frac{(i-9)\pi}{3}\right) \end{bmatrix}, \ (i = 9, 10, \ldots, 14).$$

If the state formation specified by above $h_i(t)$ $(i = 9, 10, \ldots, 14)$ is achieved, the six leaders will locate on the six vertices of a parallel hexagon respectively and keep rotation with an angular velocity of 1 rad/s. Moreover, the edge length of the desired parallel hexagon is periodic time-varying. Second, the states of the eight followers are required to converge to the convex hull formed by the states of the six leaders.

Choose $K_1 = [4, -5, -3]$ to assign the eigenvalues of $(A + BK_1)$ at $1j$, $-1j$, and -0.1 with $j^2 = -1$. In this case, the formation reference will move periodically.

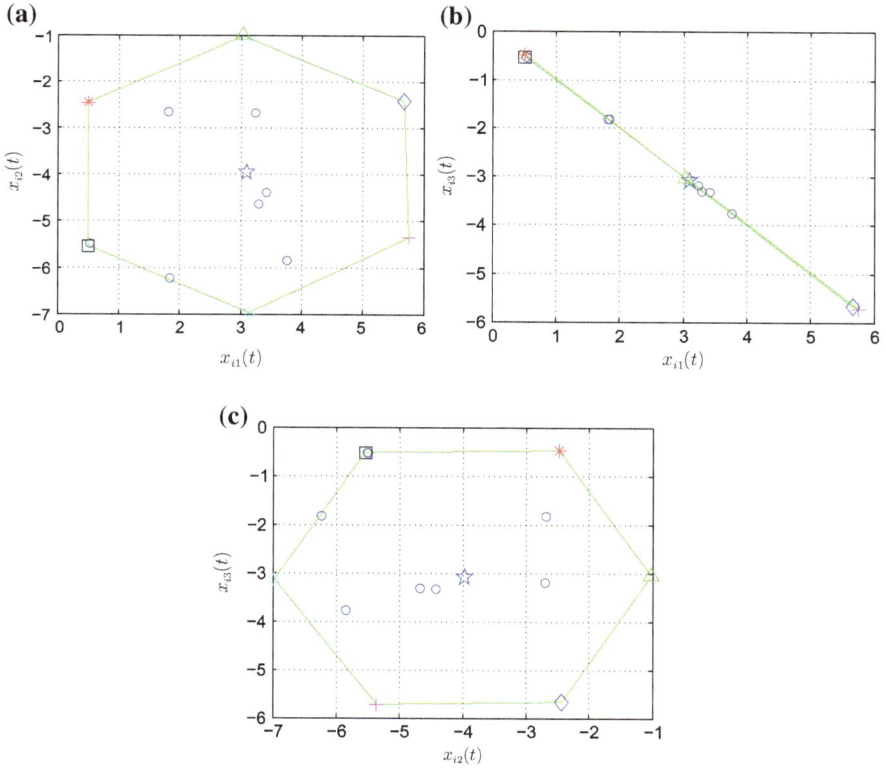

Fig. 6.3 Different views of trajectories snapshots at $t = 200$ s, **a** view of the $x_{i1}(t) - x_{i2}(t)$, **b** view of the $x_{i1}(t) - x_{i3}(t)$, **c** view of the $x_{i2}(t) - x_{i3}(t)$

It can be verified that condition (i) in Theorem 6.1 holds and (A, B) is stabilizable. Therefore, from Theorem 6.3, one can obtain matrices K_2 and K_3 to make swarm system (6.1) achieve formation-containment by protocols (6.3) and (6.4) as follows

$$K_2 = [-0.9663, 0.4601, -1.8541],\ K_3 = [-1.9326, 0.9203, -3.7083].$$

Choose the initial states of all agents as $x_{i1}(0) = (15-i)\Theta$, $x_{i2}(0) = (15-i)\Theta$, $x_{i3}(0) = (15-i)\Theta$ $(i = 1, 2, \ldots, 14)$. Figures 6.2 and 6.3 show the trajectories snapshots of the fourteen agents and the predefined formation reference function at different times, where the state trajectories of leaders are denoted by the asterisk, triangle, diamond, hexagram, plus, and square, and those of the followers and predefined formation reference function are represented by the circle and pentagram respectively. Moreover, the convex hull formed by the states of leaders is marked by solid lines. Figure 6.2a and b indicate that the leaders achieve the desired parallel hexagon formation and the states of followers converge to the convex hull formed by the states of leaders. Figure 6.2b, c and d show that the achieved formation keeps rotation around the predefined formation reference, and both the edge length of formation and the formation reference are time-varying. Figures 6.2 and 6.3 show that swarm system (6.1) achieves the desired formation-containment. It should be mentioned that in Fig. 6.3, the states of some agents are overlapped by the states of others. This is because some followers have only one neighbor, which means that when the swarm system achieves formation-containment the states of these followers will reach an agreement with those of the neighbors.

6.3 Output Formation-Containment Control

In this section, output formation-containment control problems for high-order LTI swarm systems are studied using a static output feedback control approach. Sufficient conditions for swarm systems to achieve output formation-containment are proposed and approaches to design the output formation-containment protocols are presented. It is pointed out that state/output consensus problems, state/output consensus tracking problems, state/output formation control problems, state/output containment control problems, and state formation-containment control problems can all be regarded as special cases of output formation-containment control problems.

6.3.1 Problem Description

Consider a swarm system with N agents. The dynamics of agent i is described by the following LTI system:

$$\begin{cases} \dot{x}_i(t) = Ax_i(t) + Bu_i(t), \\ y_i(t) = Cx_i(t), \end{cases} \tag{6.40}$$

where $i \in \{1, 2, \ldots, N\}$, $\mathrm{rank}(C) = q$, $x_i(t) \in \mathbb{R}^n$, $u_i(t) \in \mathbb{R}^m$, and $y_i(t) \in \mathbb{R}^q$ are the states, control inputs, and outputs of agent i, respectively. The definitions for leaders and followers, and the assumptions on the interaction topologies are the same with the ones in Sect. 6.2. A time-varying output formation for leaders is specified by a vector $h_{\mathscr{L}}(t) = [h_{M+1}^T(t), h_{M+2}^T(t), \ldots, h_N^T(t)]^T \in \mathbb{R}^{(N-M)\times q}$, where $h_i(t) \in \mathbb{R}^q$ ($i \in \mathscr{L}$) is piecewise continuously differentiable.

Definition 6.4 Swarm system (6.40) is said to *achieve output formation-containment* if for any given bounded initial states, any $i \in \mathscr{L}$ and $k \in \mathscr{F}$, there exist a vector-valued function $r(t) \in \mathbb{R}^q$ and nonnegative constants $\beta_{k,j}$ satisfying $\sum_{j=M+1}^{N} \beta_{k,j} = 1$ such that (4.37) and (5.2) hold simultaneously.

Remark 6.9 From Definitions 4.1, 4.3, 4.5, 4.7, 5.2, 6.2, 6.3 and 6.4, one sees that output formation-containment provides a unified framework for the following problems in cooperative control of swarm systems.
(i) If $C = I$, then output formation-containment problems become state formation-containment problems;
(ii) If $M = 0$ or $M = 0$ and $C = I$, then output formation-containment problems can be viewed as the time-varying output or state formation problems;
(iii) If $M = 0$ and $h_{\mathscr{L}}(t) \equiv 0$ or $M = 0$, $h_{\mathscr{L}}(t) \equiv 0$ and $C = I$, output formation-containment problems become the output or state consensus problems;
(iv) If $M = N - 1$ or $M = N - 1$ and $C = I$, then output formation-containment problems turn into the output or state consensus tracking problems;
(v) If $h_{\mathscr{L}}(t) \equiv 0$ and leaders have no neighbors, that is, for all $i, j \in \mathscr{L}$, $w_{ij} = 0$, then output formation-containment problems become the output containment problems;
(vi) If $h_{\mathscr{L}}(t) \equiv 0$, leaders have no neighbors (for all $i, j \in \mathscr{L}$, $w_{ij} = 0$) and $C = I$, then output formation-containment problems can be viewed as the state containment problems.

Figure 6.4 shows the relationship among output formation-containment control and other cooperative control problems. Moreover, for cases (i) and (ii), $r(t)$ becomes the corresponding time-varying formation reference function, and for cases (iii) and (iv), $r(t)$ becomes the corresponding time-varying consensus function.

Consider the following static output formation-containment protocols

$$u_i(t) = K_1 y_i(t) + K_2 \sum_{j \in N_i} w_{ij} \left(y_i(t) - y_j(t)\right), \ i \in \mathscr{F}, \tag{6.41}$$

$$u_i(t) = K_1 y_i(t) + K_3 \sum_{j \in N_i} w_{ij} \left((y_i(t) - h_i(t)) - (y_j(t) - h_j(t))\right), \ i \in \mathscr{L}, \tag{6.42}$$

where K_1, K_2 and K_3 are constant gain matrices with appropriate dimensions.

Remark 6.10 In protocols (6.41) and (6.42), gain matrices K_i ($i = 1, 2, 3$) can be designed in the following procedure. First, design K_1 to partially assign the motion

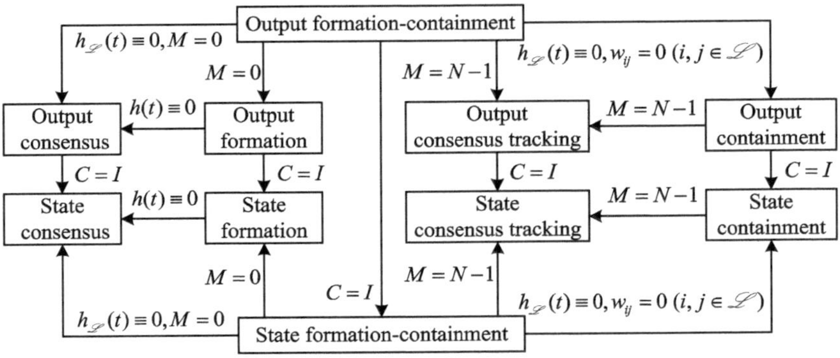

Fig. 6.4 The relationship among output formation-containment control and other cooperative control

modes of the time-varying output formation reference. Then design K_3 to ensure that the outputs of leaders achieve desired time-varying formations. Finally, design K_2 to guarantee that the outputs of followers converge to the convex hull formed by the outputs of leaders.

The current section mainly addresses the following two problems for swarm system (6.40) with protocols (6.41) and (6.42): (i) under what conditions the output formation-containment can be achieved; and (ii) how to determine the gain matrices in protocols (6.41) and (6.42) to achieve output formation-containment.

6.3.2 Problem Transformation and Preliminary Results

In this subsection, output formation-containment problems for swarm system (6.40) under protocols (6.41) and (6.42) are transformed into asymptotic stability problems and an explicit expression of the time-varying output formation reference function is given.

Under protocols (6.41) and (6.42), swarm system (6.40) can be written in a compact form as follows:

$$\begin{cases} \dot{x}_{\mathscr{F}}(t) = (I_M \otimes (A + BK_1C) + L_1 \otimes BK_2C)\, x_{\mathscr{F}}(t) + (L_2 \otimes BK_2C) x_{\mathscr{L}}(t), \\ y_{\mathscr{F}}(t) = (I_M \otimes C)\, x_{\mathscr{F}}(t), \end{cases} \tag{6.43}$$

$$\begin{cases} \dot{x}_{\mathscr{L}}(t) = (I_{N-M} \otimes (A + BK_1C) + L_3 \otimes BK_3C)\, x_{\mathscr{L}}(t) - (L_3 \otimes BK_3)\, h_{\mathscr{L}}(t), \\ y_{\mathscr{L}}(t) = (I_{N-M} \otimes C)\, x_{\mathscr{L}}(t). \end{cases} \tag{6.44}$$

Since C is of full row rank, the matrix T defined in Sect. 4.3.2 can be applied to deal with the output decomposition problems of $x_{\mathscr{F}}(t)$ and $x_{\mathscr{L}}(t)$. Applying the nonsingular transformations $I_M \otimes T$ and $I_{N-M} \otimes T$ to swarm system (6.43) and (6.44) respectively yields

$$\begin{cases} \dot{y}_{\mathscr{F}}(t) = \left(I_M \otimes (\bar{A}_{11} + \bar{B}_1 K_1) + L_1 \otimes \bar{B}_1 K_2\right) y_{\mathscr{F}}(t) + \left(I_M \otimes \bar{A}_{12}\right) \bar{y}_{\mathscr{F}}(t) \\ \qquad + (L_2 \otimes \bar{B}_1 K_2) y_{\mathscr{L}}(t), \\ \dot{\bar{y}}_{\mathscr{F}}(t) = \left(I_M \otimes (\bar{A}_{21} + \bar{B}_2 K_1) + L_1 \otimes \bar{B}_2 K_2\right) y_{\mathscr{F}}(t) + \left(I_M \otimes \bar{A}_{22}\right) \bar{y}_{\mathscr{F}}(t) \\ \qquad + (L_2 \otimes \bar{B}_2 K_2) y_{\mathscr{L}}(t), \end{cases} \tag{6.45}$$

$$\begin{cases} \dot{y}_{\mathscr{L}}(t) = \left(I_{N-M} \otimes (\bar{A}_{11} + \bar{B}_1 K_1) + L_3 \otimes \bar{B}_1 K_3\right) y_{\mathscr{L}}(t) + \left(I_{N-M} \otimes \bar{A}_{12}\right) \bar{y}_{\mathscr{L}}(t) \\ \qquad - (L_3 \otimes \bar{B}_1 K_3) h_{\mathscr{L}}(t), \\ \dot{\bar{y}}_{\mathscr{L}}(t) = \left(I_{N-M} \otimes (\bar{A}_{21} + \bar{B}_2 K_1) + L_3 \otimes \bar{B}_2 K_3\right) y_{\mathscr{L}}(t) + \left(I_{N-M} \otimes \bar{A}_{22}\right) \bar{y}_{\mathscr{L}}(t) \\ \qquad - (L_3 \otimes \bar{B}_2 K_3) h_{\mathscr{L}}(t). \end{cases} \tag{6.46}$$

Let $\tilde{y}_i(t) = y_i(t) - h_i(t)$ $(i \in \mathscr{L})$ and $\tilde{y}_{\mathscr{L}}(t) = [\tilde{y}_{M+1}^T(t), \tilde{y}_{M+2}^T(t), \ldots, \tilde{y}_N^T(t)]^T$. Then swarm system (6.46) can be rewritten as

$$\begin{cases} \dot{\tilde{y}}_{\mathscr{L}}(t) = \left(I_{N-M} \otimes \left(\bar{A}_{11} + \bar{B}_1 K_1\right) + L_3 \otimes \bar{B}_1 K_3\right) \tilde{y}_{\mathscr{L}}(t) + \left(I_{N-M} \otimes \bar{A}_{12}\right) \bar{y}_{\mathscr{L}}(t) \\ \qquad + \left(I_{N-M} \otimes \left(\bar{A}_{11} + \bar{B}_1 K_1\right)\right) h_{\mathscr{L}}(t) - \left(I_{N-M} \otimes I\right) \dot{h}_{\mathscr{L}}(t), \\ \dot{\bar{y}}_{\mathscr{L}}(t) = \left(I_{N-M} \otimes \left(\bar{A}_{21} + \bar{B}_2 K_1\right) + L_3 \otimes \bar{B}_2 K_3\right) \tilde{y}_{\mathscr{L}}(t) + \left(I_{N-M} \otimes \bar{A}_{22}\right) \bar{y}_{\mathscr{L}}(t) \\ \qquad + \left(I_{N-M} \otimes \left(\bar{A}_{21} + \bar{B}_2 K_1\right)\right) h_{\mathscr{L}}(t). \end{cases} \tag{6.47}$$

Then the following lemma holds obviously.

Lemma 6.5 *Swarm system (6.46) achieves time-varying output formation $h_{\mathscr{L}}(t)$ if and only if swarm system (6.47) achieves output consensus.*

Denote by λ_i $(i \in \mathscr{L})$ the eigenvalues of L_3, where $\lambda_{M+1} = 0$ with associated eigenvector $\bar{u}_{M+1} = \mathbf{1}/\sqrt{N-M}$ and $0 < \mathrm{Re}(\lambda_{M+2}) \le \cdots \le \mathrm{Re}(\lambda_N)$. Let $U_{\mathscr{L}}^{-1} L_3 U_{\mathscr{L}} = J_{\mathscr{L}}$, where $U_{\mathscr{L}} = [\bar{u}_{M+1}, \bar{u}_{M+2}, \ldots, \bar{u}_N]$, $U_{\mathscr{L}}^{-1} = [\tilde{u}_{M+1}, \tilde{u}_{M+2}, \ldots, \tilde{u}_N]^H$ and $J_{\mathscr{L}}$ is the Jordan canonical form of L_3. From Lemma 2.1 and the structure of $U_{\mathscr{L}}$, one can get $J_{\mathscr{L}} = \mathrm{diag}\{0, \bar{J}_{\mathscr{L}}\}$ where $\bar{J}_{\mathscr{L}}$ consists of Jordan blocks corresponding to λ_i $(i = M+2, M+3, \ldots, N)$. Let $\tilde{U}_{\mathscr{L}} = [\tilde{u}_{M+2}, \tilde{u}_{M+3}, \ldots, \tilde{u}_N]^H$, $\vartheta_{\mathscr{L}}(t) = (\tilde{u}_{M+1}^H \otimes I)\tilde{y}_{\mathscr{L}}(t)$, $\varsigma_{\mathscr{L}}(t) = (\tilde{U}_{\mathscr{L}} \otimes I)\tilde{y}_{\mathscr{L}}(t)$, $\bar{\vartheta}_{\mathscr{L}}(t) = (\tilde{u}_{M+1}^H \otimes I)\bar{y}_{\mathscr{L}}(t)$, $\bar{\varsigma}_{\mathscr{L}}(t) = [\bar{\varsigma}_{M+2}^H(t), \bar{\varsigma}_{M+3}^H(t), \ldots, \bar{\varsigma}_N^H(t)]^H = (\tilde{U}_{\mathscr{L}} \otimes I)\bar{y}_{\mathscr{L}}(t)$. Then swarm system (6.47) can be transformed into

$$\begin{cases} \dot{\vartheta}_{\mathscr{L}}(t) = \left(\bar{A}_{11} + \bar{B}_1 K_1\right) \vartheta_{\mathscr{L}}(t) + \bar{A}_{12} \bar{\vartheta}_{\mathscr{L}}(t) + \left(\tilde{u}_{M+1}^H \otimes \left(\bar{A}_{11} + \bar{B}_1 K_1\right)\right) h_{\mathscr{L}}(t) \\ \qquad - \left(\tilde{u}_{M+1}^H \otimes I\right) \dot{h}_{\mathscr{L}}(t), \\ \dot{\bar{\vartheta}}_{\mathscr{L}}(t) = \left(\bar{A}_{21} + \bar{B}_2 K_1\right) \vartheta_{\mathscr{L}}(t) + \bar{A}_{22} \bar{\vartheta}_{\mathscr{L}}(t) + \left(\tilde{u}_{M+1}^H \otimes \left(\bar{A}_{21} + \bar{B}_2 K_1\right)\right) h_{\mathscr{L}}(t), \end{cases} \tag{6.48}$$

$$\begin{cases} \dot{\varsigma}_{\mathscr{L}}(t) = \left(I_{N-M-1} \otimes \left(\bar{A}_{11} + \bar{B}_1 K_1\right) + \bar{J}_{\mathscr{L}} \otimes \bar{B}_1 K_3\right) \varsigma_{\mathscr{L}}(t) + \left(I_{N-M-1} \otimes \bar{A}_{12}\right) \bar{\varsigma}_{\mathscr{L}}(t) \\ \qquad + \left(\tilde{U}_{\mathscr{L}} \otimes \left(\bar{A}_{11} + \bar{B}_1 K_1\right)\right) h_{\mathscr{L}}(t) - \left(\tilde{U}_{\mathscr{L}} \otimes I\right) \dot{h}_{\mathscr{L}}(t), \\ \dot{\bar{\varsigma}}_{\mathscr{L}}(t) = \left(I_{N-M-1} \otimes \left(\bar{A}_{21} + \bar{B}_2 K_1\right) + \bar{J}_{\mathscr{L}} \otimes \bar{B}_2 K_3\right) \varsigma_{\mathscr{L}}(t) + \left(I_{N-M-1} \otimes \bar{A}_{22}\right) \bar{\varsigma}_{\mathscr{L}}(t) \\ \qquad + \left(\tilde{U}_{\mathscr{L}} \otimes \left(\bar{A}_{21} + \bar{B}_2 K_1\right)\right) h_{\mathscr{L}}(t). \end{cases} \tag{6.49}$$

Let

$$\tilde{y}_{\mathscr{L}C}(t) = \frac{\mathbf{1}}{\sqrt{N-M}} \otimes \vartheta_{\mathscr{L}}(t), \tag{6.50}$$

$$\tilde{y}_{\mathscr{L}\bar{C}}(t) = \tilde{y}_{\mathscr{L}}(t) - \tilde{y}_{\mathscr{L}C}(t). \tag{6.51}$$

Note that $[\vartheta_{\mathscr{L}}^H(t), 0]^H = \tilde{e}_{M+1} \otimes \vartheta_{\mathscr{L}}(t)$, where $\tilde{e}_{M+1} \in \mathbb{R}^{N-M}$ has $1/\sqrt{N-M}$ as its $i-M$th component and 0 elsewhere. One gets,

$$\tilde{y}_{\mathscr{L}C}(t) = \left(U_{\mathscr{L}} \otimes I_q\right)\left(\tilde{e}_{M+1} \otimes \vartheta_{\mathscr{L}}(t)\right) = \left(U_{\mathscr{L}} \otimes I_q\right)\left[\vartheta_{\mathscr{L}}^H(t), 0\right]^H. \tag{6.52}$$

Due to the fact that $[\vartheta_{\mathscr{L}}^H(t), \varsigma_{\mathscr{L}}^H(t)]^H = (U_{\mathscr{L}}^{-1} \otimes I_q)\tilde{y}_{\mathscr{L}}(t)$, it can be shown that

$$\tilde{y}_{\mathscr{L}\bar{C}}(t) = \tilde{y}_{\mathscr{L}}(t) - \tilde{y}_{\mathscr{L}C}(t) = \left(U_{\mathscr{L}} \otimes I_q\right)\left[0, \varsigma_{\mathscr{L}}^H(t)\right]^H. \tag{6.53}$$

Since $U_{\mathscr{L}}^{-1} \otimes I_q$ is nonsingular, from (6.52) and (6.53), one sees that $\tilde{y}_{\mathscr{L}C}(t)$ and $\tilde{y}_{\mathscr{L}\bar{C}}(t)$ are linearly independent. From (6.50) and (6.51), one can conclude that the subsystems with states $\tilde{y}_{\mathscr{L}C}(t)$ and $\tilde{y}_{\mathscr{L}\bar{C}}(t)$ describe the output consensus dynamics and disagreement output dynamics of system (6.47) respectively; that is, swarm system (6.47) achieves output consensus if and only if $\lim_{t\to\infty}\tilde{y}_{\mathscr{L}\bar{C}}(t) = 0$. Therefore, based on Lemma 6.5, the following lemma can be obtained.

Lemma 6.6 *Swarm system (6.46) achieves time-varying output formation $h_{\mathscr{L}}(t)$ if and only if*

$$\lim_{t\to\infty} \varsigma_{\mathscr{L}}(t) = 0.$$

Let

$$\psi_{\mathscr{F}i}(t) = \sum_{j\in N_i} w_{ij}\left(y_i(t) - y_j(t)\right),\ i \in \mathscr{F},$$

and $\psi_{\mathscr{F}}(t) = [\psi_{\mathscr{F}1}^T(t), \psi_{\mathscr{F}2}^T(t), \ldots, \psi_{\mathscr{F}M}^T(t)]^T$, then one has

$$\psi_{\mathscr{F}}(t) = (L_2 \otimes I)\, y_{\mathscr{L}}(t) + (L_1 \otimes I)\, y_{\mathscr{F}}(t). \tag{6.54}$$

If $\lim_{t\to\infty}\psi_{\mathscr{F}}(t) = 0$, one can obtain

$$\lim_{t\to\infty}\left(y_{\mathscr{F}}(t) - \left(-L_1^{-1}L_2 \otimes I\right) y_{\mathscr{L}}(t)\right) = 0. \tag{6.55}$$

By (6.55), Lemmas 5.1 and 6.6 the following lemma holds.

Lemma 6.7 *Swarm systems (6.40) under protocols (6.41) and (6.42) achieves output formation-containment if for any given bounded initial states*

$$\begin{cases} \lim_{t\to\infty} \varsigma_{\mathscr{L}}(t) = 0, \\ \lim_{t\to\infty} \psi_{\mathscr{F}}(t) = 0. \end{cases}$$

Based on the above analysis, an explicit expression of the time-varying output formation reference function can be presented as follows.

Lemma 6.8 *If swarm system (6.40) under protocols (6.41) and (6.42) achieves output formation-containment, then the time-varying output formation reference function $r(t)$ satisfies*

$$\lim_{t\to\infty} (r(t) - r_0(t) - r_h(t)) = 0,$$

where

$$r_0(t) = \frac{1}{\sqrt{N-M}} C e^{(A+BK_1C)t} \left(\tilde{u}_{M+1}^H \otimes I\right) x_{\mathscr{L}}(0),$$

$$r_h(t) = -\frac{1}{\sqrt{N-M}} \left(\tilde{u}_{M+1}^H \otimes I\right) h_{\mathscr{L}}(t).$$

Proof If swarm system (6.40) under protocols (6.41) and (6.42) achieves output formation-containment, then swarm system (6.46) achieves time-varying output formation $h_{\mathscr{L}}(t)$. From (6.50) to (6.53), one knows that the time-varying output formation reference function is determined by subsystem (6.48), and

$$\lim_{t\to\infty} \left(r(t) - \frac{1}{\sqrt{N-M}} \vartheta_{\mathscr{L}}(t) \right) = 0. \tag{6.56}$$

Let $\tilde{\vartheta}_{\mathscr{L}}(t) = [\vartheta_{\mathscr{L}}^H(t), \bar{\vartheta}_{\mathscr{L}}^H(t)]^H$, subsystem (6.48) can be rewritten as

$$\begin{aligned} \dot{\tilde{\vartheta}}_{\mathscr{L}}(t) = T\,(A + BK_1C)\,T^{-1}\tilde{\vartheta}_{\mathscr{L}}(t) + T \left(\tilde{u}_{M+1}^H \otimes AT^{-1} \begin{bmatrix} I \\ 0 \end{bmatrix} + \tilde{u}_{M+1}^H \otimes BK_1 \right) \\ h_{\mathscr{L}}(t) - \tilde{u}_{M+1}^H \otimes \begin{bmatrix} I \\ 0 \end{bmatrix} \dot{h}_{\mathscr{L}}(t). \end{aligned} \tag{6.57}$$

It can be verified that

$$\begin{aligned} &\int_0^t e^{(T(A+BK_1C)T^{-1})(t-\tau)} \left(\tilde{u}_{M+1}^H \otimes \begin{bmatrix} I \\ 0 \end{bmatrix} \right) \dot{h}_{\mathscr{L}}(\tau)d\tau \\ &= \left(\tilde{u}_{M+1}^H \otimes \begin{bmatrix} I \\ 0 \end{bmatrix} \right) h_{\mathscr{L}}(t) - e^{(T(A+BK_1C)T^{-1})t} \left(\tilde{u}_{M+1}^H \otimes \begin{bmatrix} I \\ 0 \end{bmatrix} \right) h_{\mathscr{L}}(0) \\ &+ \int_0^t e^{(T(A+BK_1C)T^{-1})(t-\tau)} \left(\tilde{u}_{M+1}^H \otimes (T\,(A+BK_1C)\,T^{-1}) \begin{bmatrix} I \\ 0 \end{bmatrix} \right) h_{\mathscr{L}}(\tau)d\tau, \end{aligned} \tag{6.58}$$

and $\vartheta_{\mathscr{L}}(0) = (\tilde{u}_{M+1}^H \otimes I)\,(y_{\mathscr{L}}(0) - h_{\mathscr{L}}(0))$, $\bar{\vartheta}_{\mathscr{L}}(0) = (\tilde{u}_{M+1}^H \otimes I)\bar{y}_{\mathscr{L}}(0)$. Since $y_{\mathscr{L}}(0) = (I_{N-M} \otimes C)x_{\mathscr{L}}(0)$ and $\bar{y}_{\mathscr{L}}(0) = (I_{N-M} \otimes \bar{C})x_{\mathscr{L}}(0)$, one has

$$\vartheta_{\mathscr{L}}(0) = (\tilde{u}_{M+1}^H \otimes C)x_{\mathscr{L}}(0) - (\tilde{u}_{M+1}^H \otimes I)h_{\mathscr{L}}(0), \tag{6.59}$$

$$\bar{\vartheta}_{\mathscr{L}}(0) = (\tilde{u}_{M+1}^H \otimes \bar{C})x_{\mathscr{L}}(0). \tag{6.60}$$

From (6.56) to (6.60), the conclusion of Lemma 6.8 can be obtained.

Remark 6.11 From Lemma 6.8, one sees that gain matrix K_1 can be used to partially specify the motion modes of the time-varying output formation reference. Under the cases described in Remark 6.9, corresponding formation reference function or consensus function can be obtained directly form Lemma 6.8. Moreover, if the interaction topologies are undirected, then $U_{\mathscr{L}}$ can be chosen as an orthogonal constant matrix with $\bar{u}_{M+1} = \mathbf{1}/\sqrt{N-M}$. In this case, $\tilde{u}_{M+1} = \mathbf{1}/\sqrt{N-M}$, and $r_0(t)$ and $r_h(t)$ have the following form

$$r_0(t) = Ce^{(A+BK_1C)t}\frac{1}{N-M}\sum_{i=M+1}^{N} x_i(0), \; c_h(t) = -\frac{1}{N-M}\sum_{i=M+1}^{N} h_i(t).$$

6.3.3 Output Formation-Containment Analysis and Protocol Design

In this subsection, sufficient conditions to achieve output formation-containment are proposed first. Then a method to decrease the calculation complexity of the criteria is proposed. Finally, an approach to determine the gain matrices in the protocols to achieve output formation-containment is given.

Lemma 6.7 implies that the output formation-containment of swarm system (6.40) under protocols (6.41) and (6.42) is determined by the asymptotic stability of the subsystems with state $\varsigma_{\mathscr{L}}(t)$ and output $\psi_{\mathscr{F}}(t)$ respectively. From (6.45), (6.46), (6.49), and (6.54), one sees that only observable components of $(\bar{A}_{22}, \bar{A}_{12})$ influence the stability of subsystems with state $\varsigma_{\mathscr{L}}(t)$ and output $\psi_{\mathscr{F}}(t)$. Therefore, the observability decomposition of $(\bar{A}_{22}, \bar{A}_{12})$ is presented firstly as follows. Let $\hat{y}_i(t) = \tilde{T}^{-1}\bar{y}_i(t) = [\hat{y}_{io}^T(t), \hat{y}_{i\bar{o}}^T(t)]^T$ $(i = 1, 2, \ldots, N)$, $\tilde{\varsigma}_i(t) = \tilde{T}^{-1}\bar{\varsigma}_i(t) = [\tilde{\varsigma}_{io}^T(t), \tilde{\varsigma}_{i\bar{o}}^T(t)]^T$ $(i = M+2, M+3, \ldots, N)$, $\tilde{T}^{-1}\bar{A}_{21} = [\tilde{F}_1^T, \tilde{F}_2^T]^T$, $\tilde{T}^{-1}\bar{B}_2 = [\tilde{B}_1^T, \tilde{B}_2^T]^T$, and

$$\hat{y}_{\mathscr{F}o}(t) = [\hat{y}_{1o}^T(t), \hat{y}_{2o}^T(t), \ldots, \hat{y}_{Mo}^T(t)]^T, \; \hat{y}_{F\bar{o}}(t) = [\hat{y}_{1\bar{o}}^T(t), \hat{y}_{2\bar{o}}^T(t), \ldots, \hat{y}_{M\bar{o}}^T(t)]^T,$$

$$\begin{aligned}\hat{y}_{\mathscr{L}o}(t) =& [\hat{y}_{(M+1)o}^T(t), \hat{y}_{(M+2)o}^T(t), \ldots, \hat{y}_{No}^T(t)]^T, \; \hat{y}_{\mathscr{L}\bar{o}}(t) = [\hat{y}_{(M+1)\bar{o}}^T(t),\\ &\hat{y}_{(M+2)\bar{o}}^T(t), \ldots, \hat{y}_{N\bar{o}}^T(t)]^T,\end{aligned}$$

$$\tilde{\varsigma}_{Eo}(t) = [\tilde{\varsigma}^T_{(M+2)o}(t), \tilde{\varsigma}^T_{(M+3)o}(t), \ldots, \tilde{\varsigma}^T_{No}(t)]^T, \ \tilde{\varsigma}_{\mathscr{L}\bar{o}}(t) = [\tilde{\varsigma}^T_{(M+1)\bar{o}}(t), \tilde{\varsigma}^T_{(M+2)\bar{o}}(t), \ldots, \tilde{\varsigma}^T_{N\bar{o}}(t)]^T,$$

where $\tilde{T}$ is defined in Sect. 4.3.3. Then systems (6.45), (6.46) and (6.49) can be transformed respectively into

$$\begin{cases} \dot{y}_{\mathscr{F}}(t) = \left(I_M \otimes (\bar{A}_{11} + \bar{B}_1 K_1) + L_1 \otimes \bar{B}_1 K_2\right) y_{\mathscr{F}}(t) + \left(I_M \otimes \tilde{E}_1\right) \hat{y}_{\mathscr{F}o}(t) \\ \qquad + (L_2 \otimes \bar{B}_1 K_2) y_{\mathscr{L}}(t), \\ \dot{\hat{y}}_{\mathscr{F}o}(t) = \left(I_M \otimes (\tilde{F}_1 + \tilde{B}_1 K_1) + L_1 \otimes \tilde{B}_1 K_2\right) y_{\mathscr{F}}(t) + (I_M \otimes \tilde{D}_1) \hat{y}_{\mathscr{F}o}(t) \\ \qquad + (L_2 \otimes \tilde{B}_1 K_2) y_{\mathscr{L}}(t), \\ \dot{\hat{y}}_{\mathscr{F}\bar{o}}(t) = \left(I_M \otimes (\tilde{F}_2 + \tilde{B}_2 K_1) + L_1 \otimes \tilde{B}_2 K_2\right) y_{\mathscr{F}}(t) + (I_M \otimes \tilde{D}_2) \hat{y}_{\mathscr{F}o}(t) \\ \qquad + (I_M \otimes \tilde{D}_3) \hat{y}_{\mathscr{F}\bar{o}}(t) + (L_2 \otimes \tilde{B}_2 K_2) y_{\mathscr{L}}(t), \end{cases} \tag{6.61}$$

$$\begin{cases} \dot{\varsigma}_{\mathscr{L}}(t) = \left(I_{N-M-1} \otimes (\bar{A}_{11} + \bar{B}_1 K_1) + \bar{J} \otimes \bar{B}_1 K_3\right) \varsigma_{\mathscr{L}}(t) + \left(I_{N-M-1} \otimes \tilde{E}_1\right) \tilde{\varsigma}_{\mathscr{L}o}(t) \\ \qquad + \left(\tilde{U}_{\mathscr{L}} \otimes (\bar{A}_{11} + \bar{B}_1 K_1)\right) h_{\mathscr{L}}(t) - \left(\tilde{U}_{\mathscr{L}} \otimes I\right) \dot{h}_{\mathscr{L}}(t), \\ \dot{\tilde{\varsigma}}_{\mathscr{L}o}(t) = \left(I_{N-M-1} \otimes \left(\tilde{F}_1 + \tilde{B}_1 K_1\right) + \bar{J} \otimes \tilde{B}_1 K_3\right) \varsigma_{\mathscr{L}}(t) + \left(I_{N-M-1} \otimes \tilde{D}_1\right) \tilde{\varsigma}_{\mathscr{L}o}(t) \\ \qquad + \left(\tilde{U}_{\mathscr{L}} \otimes \left(\tilde{F}_1 + \tilde{B}_1 K_1\right)\right) h_{\mathscr{L}}(t), \\ \dot{\tilde{\varsigma}}_{\mathscr{L}\bar{o}}(t) = \left(I_{N-M-1} \otimes \left(\tilde{F}_2 + \tilde{B}_2 K_1\right) + \bar{J} \otimes \tilde{B}_2 K_3\right) \varsigma_{\mathscr{L}}(t) + \left(I_{N-M-1} \otimes \tilde{D}_2\right) \tilde{\varsigma}_{\mathscr{L}o}(t) \\ \qquad + \left(I_{N-M-1} \otimes \tilde{D}_3\right) \tilde{\varsigma}_{\mathscr{L}\bar{o}}(t) + \left(\tilde{U}_{\mathscr{L}} \otimes \left(\tilde{F}_2 + \tilde{B}_2 K_1\right)\right) h_{\mathscr{L}}(t). \end{cases} \tag{6.62}$$

$$\begin{cases} \dot{\varsigma}_{\mathscr{L}}(t) = \left(I_{N-M-1} \otimes (\bar{A}_{11} + \bar{B}_1 K_1) + \bar{J} \otimes \bar{B}_1 K_3\right) \varsigma_{\mathscr{L}}(t) + \left(I_{N-M-1} \otimes \tilde{E}_1\right) \tilde{\varsigma}_{\mathscr{L}o}(t) \\ \qquad + \left(\tilde{U}_{\mathscr{L}} \otimes (\bar{A}_{11} + \bar{B}_1 K_1)\right) h_{\mathscr{L}}(t) - \left(\tilde{U}_{\mathscr{L}} \otimes I\right) \dot{h}_{\mathscr{L}}(t), \\ \dot{\tilde{\varsigma}}_{\mathscr{L}o}(t) = \left(I_{N-M-1} \otimes \left(\tilde{F}_1 + \tilde{B}_1 K_1\right) + \bar{J} \otimes \tilde{B}_1 K_3\right) \varsigma_{\mathscr{L}}(t) + \left(I_{N-M-1} \otimes \tilde{D}_1\right) \tilde{\varsigma}_{\mathscr{L}o}(t) \\ \qquad + \left(\tilde{U}_{\mathscr{L}} \otimes \left(\tilde{F}_1 + \tilde{B}_1 K_1\right)\right) h_{\mathscr{L}}(t), \\ \dot{\tilde{\varsigma}}_{\mathscr{L}\bar{o}}(t) = \left(I_{N-M-1} \otimes \left(\tilde{F}_2 + \tilde{B}_2 K_1\right) + \bar{J} \otimes \tilde{B}_2 K_3\right) \varsigma_{\mathscr{L}}(t) + \left(I_{N-M-1} \otimes \tilde{D}_2\right) \tilde{\varsigma}_{\mathscr{L}o}(t) \\ \qquad + \left(I_{N-M-1} \otimes \tilde{D}_3\right) \tilde{\varsigma}_{\mathscr{L}\bar{o}}(t) + \left(\tilde{U}_{\mathscr{L}} \otimes \left(\tilde{F}_2 + \tilde{B}_2 K_1\right)\right) h_{\mathscr{L}}(t). \end{cases} \tag{6.63}$$

The following theorem presents sufficient conditions for swarm systems to achieve output formation-containment.

Theorem 6.4 *For any bounded initial states, swarm system (6.40) under protocols (6.41) and (6.42) achieves output formation-containment if the following conditions hold simultaneously*

(i) For all $i \in \mathscr{L}$

$$\lim_{t\to\infty}\left(\begin{bmatrix}\bar{A}_{11}+\bar{B}_1K_1\\ \tilde{F}_1+\tilde{B}_1K_1\end{bmatrix}\left(h_i(t)-h_j(t)\right)-\begin{bmatrix}I\\0\end{bmatrix}\left(\dot{h}_i(t)-\dot{h}_j(t)\right)\right)=0,\ j\in N_i; \tag{6.64}$$

(ii) For all $i \in \mathscr{F}$, $\Phi_{\mathscr{F}i}$ are Hurwitz, where

$$\Phi_{\mathscr{F}i}=\begin{bmatrix}\bar{A}_{11}+\bar{B}_1K_1+\lambda_i\bar{B}_1K_2 & \tilde{E}_1\\ \tilde{F}_1+\tilde{B}_1K_1+\lambda_i\tilde{B}_1K_2 & \tilde{D}_1\end{bmatrix};$$

(iii) For all $i \in \{M+2, M+3, \ldots, N\}$, $\Phi_{\mathscr{L}i}$ are Hurwitz, where

$$\Phi_{\mathscr{L}i}=\begin{bmatrix}\bar{A}_{11}+\bar{B}_1K_1+\lambda_i\bar{B}_1K_3 & \tilde{E}_1\\ \tilde{F}_1+\tilde{B}_1K_1+\lambda_i\tilde{B}_1K_3 & \tilde{D}_1\end{bmatrix}.$$

Proof Due to the fact that $\tilde{U}_{\mathscr{L}}\mathbf{1}=0$, if condition (i) holds, one has

$$\lim_{t\to\infty}\left(\left(\tilde{U}_{\mathscr{L}}\otimes\begin{bmatrix}\bar{A}_{11}+\bar{B}_1K_1\\ \tilde{F}_1+\tilde{B}_1K_1\end{bmatrix}\right)h_{\mathscr{L}}(t)-\left(\tilde{U}_{\mathscr{L}}\otimes\begin{bmatrix}I\\0\end{bmatrix}\right)\dot{h}_{\mathscr{L}}(t)\right)=0. \tag{6.65}$$

Consider the stabilities of the following $N-M-1$ subsystems

$$\dot{\rho}_i(t)=\begin{bmatrix}\bar{A}_{11}+\bar{B}_1K_1+\lambda_i\bar{B}_1K_3 & \tilde{E}_1\\ \tilde{F}_1+\tilde{B}_1K_1+\lambda_i\tilde{B}_1K_3 & \tilde{D}_1\end{bmatrix}\rho_i(t)\ (i=M+2,M+3,\ldots,N). \tag{6.66}$$

If condition (iii) holds, then the $N-M-1$ subsystems described by (6.66) are asymptotically stable. By the structure of $\bar{J}_{\mathscr{L}}$, it holds that the system described by

$$\dot{\bar{\rho}}(t)=\begin{bmatrix}I_{N-M-1}\otimes\left(\bar{A}_{11}+\bar{B}_1K_1\right)+\bar{J}\otimes\bar{B}_1K_3 & I_{N-M-1}\otimes\tilde{E}_1\\ I_{N-M-1}\otimes\left(\tilde{F}_1+\tilde{B}_1K_1\right)+\bar{J}\otimes\tilde{B}_1K_3 & I_{N-M-1}\otimes\tilde{D}_1\end{bmatrix}\bar{\rho}(t), \tag{6.67}$$

is asymptotically stable. From (6.63), (6.65) and (6.67), one knows that

$$\lim_{t\to\infty}\varsigma_{\mathscr{L}}(t)=0. \tag{6.68}$$

By (6.68) and Lemma 6.6, swarm system (6.46) achieves time-varying output formation $h_{\mathscr{L}}(t)$.

When leaders achieve time-varying output formation $h_{\mathscr{L}}(t)$, then

$$\begin{cases}\lim_{t\to\infty}\left((L_2L_3)\otimes(\bar{B}_1K_3)\right)\left(y_{\mathscr{L}}(t)-h_{\mathscr{L}}(t)\right)=\lim_{t\to\infty}\left((L_2L_3)\otimes(\bar{B}_1K_3)\right)(\mathbf{1}\otimes r(t)),\\ \lim_{t\to\infty}\left((L_2L_3)\otimes(\tilde{B}_1K_3)\right)\left(y_{\mathscr{L}}(t)-h_{\mathscr{L}}(t)\right)=\lim_{t\to\infty}\left((L_2L_3)\otimes(\tilde{B}_1K_3)\right)(\mathbf{1}\otimes r(t)).\end{cases} \tag{6.69}$$

Note that $L_3\mathbf{1} = 0$. One has

$$\begin{cases} \lim_{t\to\infty} \left((L_2L_3)\otimes(\bar{B}_1K_3)\right)(y_{\mathcal{L}}(t) - h_{\mathcal{L}}(t)) = 0, \\ \lim_{t\to\infty} \left((L_2L_3)\otimes(\tilde{B}_1K_3)\right)(y_{\mathcal{L}}(t) - h_{\mathcal{L}}(t)) = 0. \end{cases} \tag{6.70}$$

Let

$$\tilde{\varphi}_{\mathcal{F}i}(t) = \sum_{j\in N_i} w_{ij}\left(\hat{y}_{io}(t) - \hat{y}_{jo}(t)\right),\ i \in \mathcal{F},$$

and $\tilde{\varphi}_{\mathcal{F}}(t) = [\tilde{\varphi}^T_{\mathcal{F}1}(t), \tilde{\varphi}^T_{\mathcal{F}2}(t), \ldots, \tilde{\varphi}^T_{\mathcal{F}M}(t)]^T$. Then one has

$$\tilde{\varphi}_{\mathcal{F}}(t) = (L_2\otimes I)\,\hat{y}_{\mathcal{L}o}(t) + (L_1\otimes I)\,\hat{y}_{\mathcal{F}o}(t). \tag{6.71}$$

Taking the derivative of (6.54) and (6.71) with respect to t respectively, one can obtain

$$\begin{cases} \dot{\psi}_{\mathcal{F}}(t) = (L_2\otimes I)\,\dot{y}_{\mathcal{L}}(t) + (L_1\otimes I)\,\dot{y}_{\mathcal{F}}(t), \\ \dot{\tilde{\varphi}}_{\mathcal{F}}(t) = (L_2\otimes I)\,\dot{\hat{y}}_{\mathcal{L}o}(t) + (L_1\otimes I)\,\dot{\hat{y}}_{\mathcal{F}o}(t). \end{cases} \tag{6.72}$$

From (6.54), (6.61), (6.62), and (6.70)–(6.72), one has

$$\begin{bmatrix} \dot{\psi}_{\mathcal{F}}(t) \\ \dot{\tilde{\varphi}}_{\mathcal{F}}(t) \end{bmatrix} = \begin{bmatrix} I_M\otimes(\bar{A}_{11} + \bar{B}_1K_1) + L_1\otimes\bar{B}_1K_2 & I_M\otimes\tilde{E}_1 \\ I_M\otimes(\tilde{F}_1 + \tilde{B}_1K_1) + L_1\otimes\tilde{B}_1K_2 & I_M\otimes\tilde{D}_1 \end{bmatrix}\begin{bmatrix} \psi_{\mathcal{F}}(t) \\ \tilde{\varphi}_{\mathcal{F}}(t) \end{bmatrix}. \tag{6.73}$$

If condition (ii) holds, then the following M subsystems

$$\dot{\tilde{\rho}}_i(t) = \begin{bmatrix} \bar{A}_{11} + \bar{B}_1K_1 + \lambda_i\bar{B}_1K_2 & \tilde{E}_1 \\ \tilde{F}_1 + \tilde{B}_1K_1 + \lambda_i\tilde{B}_1K_2 & \tilde{D}_1 \end{bmatrix}\tilde{\rho}_i(t)\ (i = 1, 2, \ldots, M), \tag{6.74}$$

are asymptotically stable. From (6.74), and the structure of $U_{\mathcal{F}}$ and $\bar{\Lambda}_{\mathcal{F}}$, one knows that the system described by (6.73) is asymptotically stable. Therefore,

$$\lim_{t\to\infty} \psi_{\mathcal{F}}(t) = 0. \tag{6.75}$$

From (6.68), (6.75), and Lemma 6.7, one sees that swarm system (6.40) under protocols (6.41) and (6.42) achieves output formation-containment. The proof of Theorem 6.4 is completed.

It should be pointed out that there exist $N - 1$ Hurwitz constraints in Theorem 6.4. The following theorem presents an approach to decrease the calculation complexity.

For simplicity of expression, let

$$\bar{A}_o = \begin{bmatrix} \bar{A}_{11} + \bar{B}_1K_1 & \tilde{E}_1 \\ \tilde{F}_1 + \tilde{B}_1K_1 & \tilde{D}_1 \end{bmatrix},\ B_o = \begin{bmatrix} \bar{B}_1 \\ \tilde{B}_1 \end{bmatrix},\ C_o = \begin{bmatrix} I & 0 \end{bmatrix},$$

then $\Phi_{\mathscr{F}i}$ $(i \in \mathscr{F})$ and $\Phi_{\mathscr{L}k}$ $(k = M+2, M+3, \ldots, N)$ can be rewritten as

$$\Phi_{\mathscr{F}i} = \bar{A}_o + \lambda_i B_o K_2 C_o \ (i \in \mathscr{F}), \tag{6.76}$$

$$\Phi_{\mathscr{L}k} = \bar{A}_o + \lambda_k B_o K_3 C_o \ (k = M+2, M+3, \ldots, N). \tag{6.77}$$

From (6.76) and (6.77), one sees that $\Phi_{\mathscr{F}i}$ and $\Phi_{\mathscr{L}k}$ are Hurwitz if and only if subsystems $(\bar{A}_o, \lambda_i B_o, C_o)$ and $(\bar{A}_o, \lambda_k B_o, C_o)$ can be stabilized by the static output feedback controllers with gains K_2 and K_3, respectively. Based on the decomposition of the real and imaginary parts, one knows that $\Phi_{\mathscr{F}i}$ and $\Phi_{\mathscr{L}k}$ are Hurwitz if and only if that subsystems $(\Lambda_{\bar{A}_o}, \Psi_{\lambda_i}\Lambda_{B_o}, \Lambda_{C_o})$ and $(\Lambda_{\bar{A}_o}, \Psi_{\lambda_k}\Lambda_{B_o}, \Lambda_{C_o})$ can be stabilized by the static output feedback controllers with gains Λ_{K_2} and Λ_{K_3}, respectively.

Theorem 6.5 *For any bounded initial states, swarm system (6.40) under protocols (6.41) and (6.42) achieves output formation-containment if the condition (i) in Theorem 6.4 holds, and there exist real matrices K_2, K_3, $\tilde{R}_{\mathscr{F}} = \tilde{R}_{\mathscr{F}}^T > 0$ and $\tilde{R}_{\mathscr{L}} = \tilde{R}_{\mathscr{L}}^T > 0$ such that*

$$(\Lambda_{\bar{A}_o} + \Psi_{\bar{\lambda}_i}\Lambda_{B_o}\Lambda_{K_2}\Lambda_{C_o})^T \tilde{R}_{\mathscr{F}} + \tilde{R}_{\mathscr{F}}(\Lambda_{\bar{A}_o} + \Psi_{\bar{\lambda}_i}\Lambda_{B_o}\Lambda_{K_2}\Lambda_{C_o}) < 0 \quad (i = 1, 2, 3, 4), \tag{6.78}$$

$$(\Lambda_{\bar{A}_o} + \Psi_{\tilde{\lambda}_i}\Lambda_{B_o}\Lambda_{K_3}\Lambda_{C_o})^T \tilde{R}_{\mathscr{L}} + \tilde{R}_{\mathscr{L}}(\Lambda_{\bar{A}_o} + \Psi_{\tilde{\lambda}_i}\Lambda_{B_o}\Lambda_{K_3}\Lambda_{C_o}) < 0 \quad (i = 1, 2, 3, 4). \tag{6.79}$$

Proof It will be shown that conditions (6.78) and (6.79) can make sure that conditions (ii) and (iii) in Theorem 6.4 hold, respectively. First, define $X_i = (\Lambda_{\bar{A}_o} + \Psi_{\lambda_i}\Lambda_{B_o}\Lambda_{K_2}\Lambda_{C_o})^T \tilde{R}_{\mathscr{F}} + \tilde{R}_{\mathscr{F}}(\Lambda_{\bar{A}_o} + \Psi_{\lambda_i}\Lambda_{B_o}\Lambda_{K_2}\Lambda_{C_o})$ $(i \in \mathscr{F})$, then there exist real symmetric matrices $\tilde{\Omega}_{\mathscr{F}0}$, $\tilde{\Omega}_{\mathscr{F}1}$ and $\tilde{\Omega}_{\mathscr{F}2}$ independent of λ_i $(i = 1, 2, \ldots, N)$, $\bar{\lambda}_i$ and $\tilde{\lambda}_i$ $(i = 1, 2, 3, 4)$ such that X_i can be written as

$$X_i = \tilde{\Omega}_{\mathscr{F}0} + \mathrm{Re}(\lambda_i)\tilde{\Omega}_{\mathscr{F}1} + \mathrm{Im}(\lambda_i)\tilde{\Omega}_{\mathscr{F}2}. \tag{6.80}$$

Similarly, define $Z_i = (\Lambda_{\bar{A}_o} + \Psi_{\bar{\lambda}_i}\Lambda_{B_o}\Lambda_{K_2}\Lambda_{C_o})^T \tilde{R}_{\mathscr{F}} + \tilde{R}_{\mathscr{F}}(\Lambda_{\bar{A}_o} + \Psi_{\bar{\lambda}_i}\Lambda_{B_o}\Lambda_{K_2}\Lambda_{C_o})$ $(i = 1, 2, 3, 4)$, then it holds

$$Z_i = \tilde{\Omega}_{\mathscr{F}0} + \mathrm{Re}\left(\bar{\lambda}_i\right)\tilde{\Omega}_{\mathscr{F}1} + \mathrm{Im}\left(\bar{\lambda}_i\right)\tilde{\Omega}_{\mathscr{F}2}. \tag{6.81}$$

By (6.80), (6.81), and Lemma 4.3, one has that if condition (6.78) holds, then $X_i < 0$ $(i \in \mathscr{F})$ holds.

Consider the stability of the following subsystem

$$\dot{\zeta}_i(t) = \left(\Lambda_{\bar{A}_o} + \Psi_{\lambda_i}\Lambda_{B_o}\Lambda_{K_2}\Lambda_{C_o}\right)\zeta_i(t), \ i \in \mathscr{F}. \tag{6.82}$$

Choose a Lyapunov functional candidate as follows:

$$V_{\mathscr{F}i}(t) = \tilde{\zeta}_i^T(t)\tilde{R}_{\mathscr{F}}\tilde{\zeta}_i(t). \tag{6.83}$$

Taking the time derivative of $V_{\mathscr{F}i}(t)$ along the trajectory of (6.82), one can obtain

$$\begin{aligned}\dot{V}_{\mathscr{F}i}(t) &= \tilde{\zeta}_i^T(t)(\Lambda_{\bar{A}_o} + \Psi_{\lambda_i}\Lambda_{B_o}\Lambda_{K_2}\Lambda_{C_o})^T\tilde{R}_{\mathscr{F}}\tilde{\zeta}_i(t) \\ &\quad + \tilde{\zeta}_i^T(t)\tilde{R}_{\mathscr{F}}(\Lambda_{\bar{A}_o} + \Psi_{\lambda_i}\Lambda_{B_o}\Lambda_{K_2}\Lambda_{C_o})\tilde{\zeta}_i(t).\end{aligned}$$

Note that $X_i < 0$ ($i \in \mathscr{F}$), then $\dot{V}_{\mathscr{F}i}(t) < 0$; that is, $\Phi_{\mathscr{F}i}$ ($i \in \mathscr{F}$) is Hurwitz. By a similar analysis as for $\Phi_{\mathscr{F}i}$, it can be obtained that condition (6.79) can make sure that $\Phi_{\mathscr{L}k}$ ($k = M+2, M+3, \ldots, N$) are Hurwitz. Therefore, from Theorem 6.4, the conclusion of Theorem 6.5 can be proofed.

Remark 6.12 Theorem 6.5 decreases the calculation complexity of Theorem 6.4 by reducing the simultaneous stabilization problems of $N-1$ subsystems into that of eight. Since the conditions (6.78) and (6.79) in Theorem 6.5 are independent of the number of agents. As a result, for high-order LTI swarm systems with directed interaction topologies and a huge number of agents, the calculation complexity can be decreased significantly, although it may bring some conservation.

In the following, the algorithm proposed by He and Wang [17] is applied to solve the eight quadratic matrix inequality (QMI) constraints in Theorem 6.5 for K_2 and K_3. It should be pointed out that algorithm to solve the QMI has been given in Sect. 4.3.3. For simplicity of description, only the necessary condition for the convergence of the algorithm is presented in this subsection and the explicit expression of the algorithm can refer to Sect. 4.3.3.

Using the PBH criteria for stabilizability and observability, the following lemma can be easily verified.

Lemma 6.9 *If* (A, B) *is stabilizable, then for any* $\lambda \in \mathbb{C}$ *with* $\mathrm{Re}(\lambda) > 0$, $(\Lambda_{\bar{A}_o}, \Upsilon_\lambda\Lambda_{B_o})$ *is stabilizable. If* $(\tilde{D}_1, \tilde{E}_1)$ *is completely observable, then* $(\Lambda_{\bar{A}_o}, \Lambda_{C_o})$ *is detectable.*

From Lemma 6.9, one knows that if (A, B) is stabilizable, then $(\Lambda_{\bar{A}_o}, \Psi_{\bar{\lambda}_i}\Lambda_{B_o})$ ($i = 1, 2, 3, 4$) and $(\Lambda_{\bar{A}_o}, \Psi_{\tilde{\lambda}_i}\Lambda_{B_o})$ ($i = 1, 2, 3, 4$) are stabilizable. Moreover, since $(\tilde{D}_1, \tilde{E}_1)$ is completely observable, then $(\Lambda_{\bar{A}_o}, \Lambda_{C_o})$ is detectable. According to the results in He and Wang [17], the convergence of the algorithm can be guaranteed. The QMIs (6.78) and (6.79) in Theorem 6.5 have the same structure as the QMI (4.71) in Theorem 4.7. Therefore, by direct variable substitution, Algorithm 4.2 can be applied to solve QMIs (6.78) and (6.79) for K_2 and K_3.

6.3.4 Numerical Simulations

In this section, a numerical example is given to demonstrate theoretical results obtained in this section.

Consider a sixth-order swarm system with six followers and six leaders. The interaction topology of the swarm system in the example is shown in Fig. 6.5. The dynamics of each agent is described by (6.40) with $x_i(t) = [x_{i1}(t), x_{i2}(t), \ldots, x_{i6}(t)]^T$, $y_i(t) = [y_{i1}(t), y_{i2}(t), y_{i3}(t)]^T$ $(i = 1, 2, \ldots, 12)$, and

$$A = \begin{bmatrix} -3.25 & -2.5 & -1 & 1.25 & -1.75 & -1.5 \\ -1.5 & 2 & -1 & -1.5 & 4.5 & 2 \\ -4.25 & 1.5 & 0 & -1.75 & 4.25 & 2.5 \\ 8.5 & 2 & 1 & 0.5 & -3.5 & 0 \\ 3.25 & -2.5 & 2 & 1.75 & -5.25 & -1.5 \\ 3 & 0 & -3 & 2 & -4 & -3 \end{bmatrix}, \; B = \begin{bmatrix} 1 \\ 2 \\ 1 \\ 0 \\ 1 \\ 0 \end{bmatrix}, \; C = \begin{bmatrix} 0 & 0 & 1 & 0 & 1 & 0 \\ 1 & 0 & 0 & 1 & -1 & 0 \\ 0 & 1 & 0 & -1 & 0 & 1 \end{bmatrix}.$$

Choose

$$\bar{C} = \begin{bmatrix} 1 & 0 & -1 & 0 & 0 & 0 \\ 0 & 1 & 0 & 1 & 0 & -1 \\ 0 & 1 & -1 & 0 & 1 & 0 \end{bmatrix}.$$

It can be shown that (A, B) is stabilizable and $(\bar{A}_{22}, \bar{A}_{12})$ is not completely observable. Choose a nonsingular matrix $\tilde{T}$ as follows:

$$\tilde{T} = \begin{bmatrix} 1 & 0 & -1 \\ 0 & -1 & 2 \\ 1 & -1 & 0 \end{bmatrix}.$$

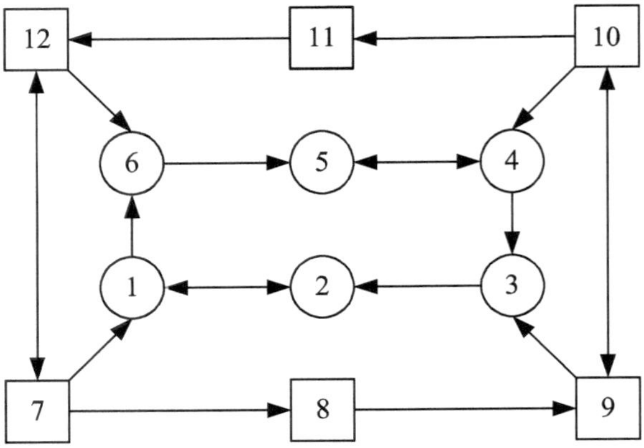

Fig. 6.5 Directed interaction topology G

The outputs of leaders are required to preserve a periodic time-varying parallel hexagon formation and keep rotation around the predefined time-varying output formation reference. The time-varying output formation for leaders is defined as follows:

$$h_i(t) = \begin{bmatrix} 6\cos\left(t + \frac{(i-7)\pi}{3}\right) \\ -6\sin\left(t + \frac{(i-7)\pi}{3}\right) \\ -6\cos\left(t + \frac{(i-7)\pi}{3}\right) \end{bmatrix} \quad (i = 7, 8, \ldots, 12).$$

If the time-varying output formation specified by the above $h_i(t)$ $(i = 7, 8, \ldots, 12)$ is achieved, the outputs of six leaders will locate on the six vertices of a parallel hexagon respectively and keep rotation with an angular velocity of 1rad/s. In addition, the edge length of the desired parallel hexagon is periodic time-varying.

Choose $K_1 = [-2, 0, -2]$ to assign the eigenvalues of $A + BK_1C$ at $1j$, $-1j$, $-1.7402 + 0.9370j$, $-1.7402 - 0.9370j$, -2, and -11.5195, where $j^2 = -1$. In this configuration, the time-varying output formation can move periodically. It can

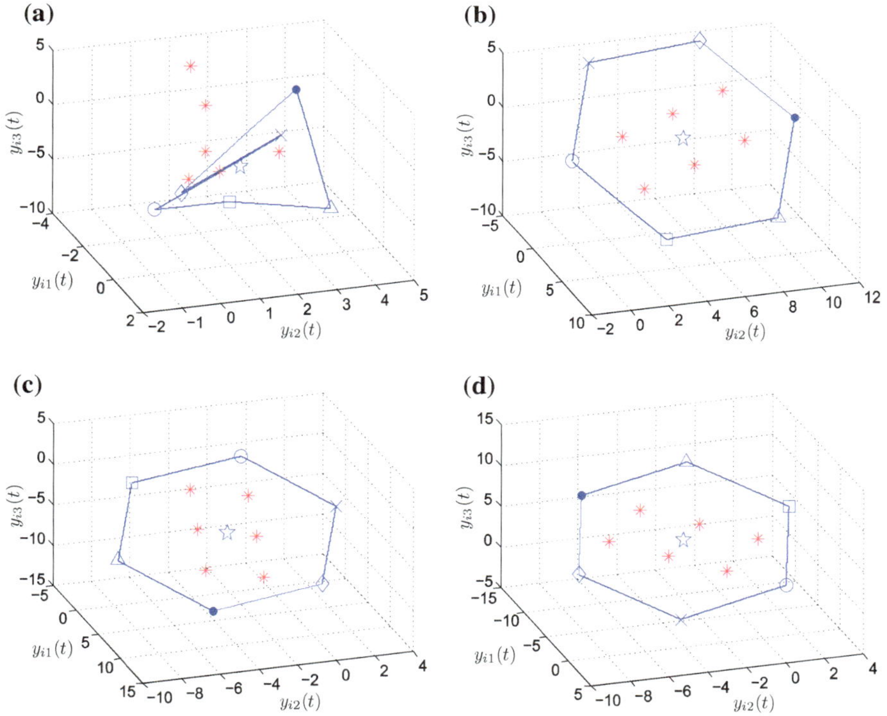

Fig. 6.6 Output snapshots of twelve agents and $r(t)$. **a** $t = 0$ s, **b** $t = 146$ s, **c** $t = 148$ s, **d** $t = 150$ s

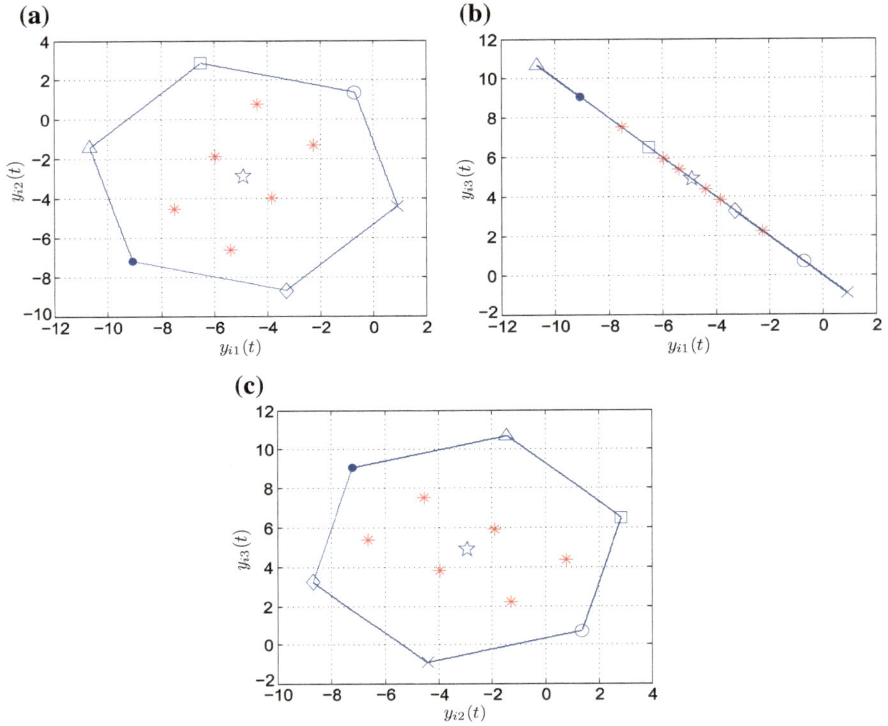

Fig. 6.7 Different views of output snapshots and $r(t)$ at $t = 150$ s, **a** view of $y_{i1}(t) - y_{i2}(t)$, **b** view of $y_{i1}(t) - y_{i3}(t)$, **c** view of $y_{i2}(t) - y_{i3}(t)$

be verified that condition (i) in Theorem 6.4 is satisfied. Using the similar algorithm as Algorithm 4.2 to solve QMIs (6.78) and (6.79), one can obtain K_2 and K_3 as

$$K_2 = [-3.0625, 0.1068, -1.7663], \ K_3 = [-2.6361, 0.1878, -1.3892].$$

Let the initial states of each agent be $x_{ij}(0) = i\,(\Theta - 0.5)$ $(i = 1, 2, \ldots, 12;\ j = 1, 2, \ldots, 6)$.

Figure 6.6 shows the output snapshots of the twelve agents and the predefined time-varying output formation reference function $r(t)$ at different times, where the outputs of leaders are denoted by the point, triangle, square, circle, x-mark, and diamond, and the outputs of the followers and predefined output formation reference function are represented by the asterisk and pentagram. The convex hull formed by the outputs of leaders is depicted by solid lines. Figure 6.7 illustrates different views of output snapshots and $r(t)$ at $t = 150$s. Figure 6.6a and b indicate that the outputs of leaders achieve a parallel hexagon formation and the point corresponding to the predefined formation reference function $r(t)$ lies in the center of the formation. Figure 6.6b, c and d show that the achieved output formation keeps rotation around the predefined output formation reference, and both the parallel hexagon formation and the output formation reference are time-varying. Therefore, the desired time-varying output formation is achieved by leaders. Figures 6.6b, c, d and 6.7 reveal that the outputs of followers converge to the convex hull formed by the outputs of leaders. Therefore, the output formation-containment is achieved.

6.4 Conclusions

In this chapter, state and output formation-containment control problems for high-order LTI swarm systems were studied, respectively. Sufficient conditions for swarm systems to achieve state and output formation-containment were proposed. Necessary and sufficient conditions for swarm systems to achieve time-varying state formation and state containment were presented as special cases. Explicit expressions of state and output formation reference functions were given and approaches to design the state and output formation-containment protocols were proposed. It was shown that state/output consensus problems, state/output consensus tracking problems, state/output formation control problems, state/output containment control problems, and state formation-containment problems can all be treated as special cases of output formation-containment control problems. The results in this chapter are mainly based on [18].

References

1. Ji M, Ferrari-Trecate G, Egerstedt M et al (2008) Containment control in mobile networks. IEEE Trans Autom Control 53(8):1972–1975
2. Meng ZY, Ren W, You Z (2010) Distributed finite-time attitude containment control for multiple rigid bodies. Automatica 46(12):2092–2099
3. Notarstefano G, Egerstedt M, Haque M (2011) Containment in leader-follower networks with switching communication topologies. Automatica 47(5):1035–1040
4. Cao YC, Ren W, Egerstedt M (2012) Distributed containment control with multiple stationary or dynamic leaders in fixed and switching directed networks. Automatica 48(8):1586–1597
5. Liu HY, Xie GM, Wang L (2012) Necessary and sufficient conditions for containment control of networked multi-agent systems. Automatica 48(7):1415–1422
6. Lou YC, Hong YG (2012) Target containment control of multi-agent systems with random switching interconnection topologies. Automatica 48(5):879–885
7. Liu HY, Xie GM, Wang L (2012) Containment of linear multi-agent systems under general interaction topologies. Syst Control Lett 61(4):528–534
8. Li ZK, Ren W, Liu XD et al (2013) Distributed containment control of multi-agent systems with general linear dynamics in the presence of multiple leaders. Int J Robust Nonlinear Control 23(5):534–547
9. Dong XW, Xi JX, Lu G et al (2014) Containment analysis and design for high-order linear time-invariant singular swarm systems with time delays. Int J Robust Nonlinear Control 24(7):1189–1204
10. Dong XW, Shi ZY, Lu G et al (2015) Output containment analysis and design for high-order linear time-invariant swarm systems. Int J Robust Nonlinear Control 25(6):900–913

11. Ferrari-Trecate G, Egerstedt M, Buffa A, et al (2006) Laplacian sheep: a hybrid, stop-go policy for leader-based containment control. In: Proceedings of Hybrid Systems: Computation and Control, pp 212–226
12. Dimarogonas DV, Egerstedt M, Kyriakopoulos KJ (2006) A leader-based containment control strategy for multiple unicycles. In: Proceedings of the 45th IEEE Conference on Decision and Control, pp 5968–5973
13. Xi JX, Cai N, Zhong YS (2010) Consensus problems for high-order linear time-invariant swarm systems. Physica A 389(24):5619–5627
14. Ni W, Cheng DZ (2010) Leader-following consensus of multi-agent systems under fixed and switching topologies. Syst Control Lett 59(3–4):209–217
15. Lafferriere G, Williams A, Caughman J et al (2005) Decentralized control of vehicle formations. Syst Control Lett 54(9):899–910
16. Ma CQ, Zhang JF (2012) On formability of linear continuous-time multi-agent systems. J Syst Sci Complex 25(1):13–29
17. He Y, Wang Q (2006) An improved ILMI method for static output feedback control with application to multivariable PID control. IEEE Trans Autom Control 51(10):1678–1683
18. Dong XW, Shi ZY, Lu G, et al (2014) Formation-containment analysis and design for high-order linear time-invariant swarm systems. Int J Robust Nonlinear Control, in press. doi:10.1002/rnc.3274

Chapter 7
Conclusions and Future Work

Abstract In this chapter, the whole work of this thesis is concluded, and some interesting and challenging problems on formation and containment control of high-order swarm systems are pointed out as the future work.

7.1 Conclusions

In this thesis, formation and containment control problems for high-order linear time-invariant (LTI) swarm systems with directed interaction topologies were studied. Four related research topics in cooperative control of swarm systems, that is, consensus control, formation control, containment control, and formation-containment control, were discussed, respectively. Moreover, the theoretical results on time-varying formation control of high-order swarm systems were applied to deal with the time-varying formation control problems of unmanned aerial vehicle (UAV) swarm systems, and time-varying formation flight experiments were presented to demonstrate the theoretical results. The main contribution of this thesis is summarized as follows.

(1) Consensus control of swarm systems
For high-order LTI swarm systems with nonuniform time-varying delays, interaction uncertainties, and time-varying external disturbances belonging to L_2 or L_∞, practical consensus problems were proposed and studied. A practical consensus protocol was constructed using dynamic output feedback approaches. Based on state space decomposition approaches, practical consensus problems of swarm systems were transformed into stability problems of disagreement subsystems. Sufficient conditions for swarm systems to achieve practical consensus were presented using Lyapunov-Krasovskii functional approaches and linear matrix inequality techniques. Explicit expressions of the practical consensus function and practical consensus error bounds were derived. The results can also be applied to solve the consensus problems of swarm systems with directed interaction topologies where time delays, interaction uncertainties, and external disturbances partially exist.

X. Dong, *Formation and Containment Control for High-order Linear Swarm Systems*, Springer Theses, DOI 10.1007/978-3-662-47836-3_7

(2) Formation control of swarm systems
For time-varying state formation control problems of high-order LTI swarm systems, a formation protocol with time delays was constructed. It was pointed out that many existing consensus-based formation control protocols can be regarded as special cases of the one in this thesis. By state transformation and state space decomposition, time-varying state formation problems of swarm systems were converted into the stability problem of time-delayed system. Necessary and sufficient conditions for swarm systems with time delays to achieve time-varying state formations were proposed, and necessary and sufficient conditions for time-varying formation feasibilities were presented. An explicit expression of the state formation reference function was derived, and approaches to assign the motion modes of the state formation reference were given. Approaches to expand the feasible formation set and design the time-delayed state formation protocol were proposed. All the results on time-varying state formation control can be used to deal with time-varying/time-invariant state formation control problems and state consensus control problems of swarm systems with/with out time delays. For time-varying output formation control problems of high-order LTI swarm systems, a static output formation protocol was constructed using the outputs of neighboring agents. Using the coordinate transformation, output space decomposition and partial stability theory, necessary and sufficient conditions for swarm systems to achieve time-varying output formations, and output formation feasibilities were presented. Explicit expressions of the output formation reference function, and approaches to partially specify motion modes of the output formation reference and expand the feasible output formation set were proposed. An algorithm to design the output formation protocol was provided. It was pointed out that the results on time-varying output formation control can be applied to solve the state/output consensus problems of swarm systems. Moreover, theories on time-varying formations of high-order swarm systems were applied to deal with the time-varying formation control problems of UAV swarm systems. Necessary and sufficient conditions for UAV swarm systems to achieve time-varying formations and approaches to design the formation protocol were derived. It was proved that the collisions among UAVs can be avoided during the formation control by appropriately choosing the initial states of each UAV. Series of time-varying formation control experiments were performed using five quadrotor UAVs to demonstrate the effectiveness of the theoretical results in outdoor environment.

(3) Containment control of swarm systems
For output containment control problems of high-order LTI swarm system, an output containment protocol was proposed based on dynamic output feedback control. By output decomposition, observability decomposition, and coordinate transformation, necessary and sufficient conditions for swarm systems to achieve output containment were presented. Approaches to partially assign the motion modes of leaders and design the dynamic output containment protocol were given. For state containment control problems of high-order LTI singular swarm systems with time delays, a state containment was constructed. By model transformation, containment problems of

singular swarm systems were converted into stability problems of multiple low-dimensional time-delayed systems. In terms of linear matrix inequalities, sufficient conditions were presented for time-delayed singular swarm systems to achieve state containment, which were independent of the number of agents. By using the method of changing variables, an approach was provided to determine the gain matrices in the protocols. Results on state containment of singular swarm systems with time delays can be applied to deal with the consensus tracking control problems for singular swarm systems with time delays, state consensus tracking control problems, and state containment control problems of time-delayed swarm systems.

(4) Formation-containment control of swarm systems
State formation-containment analysis and design problems for high-order LTI swarm systems were dealt first. State formation-containment protocols were presented for leaders and followers respectively to drive the states of leaders to realize the predefined time-varying formation and propel the states of followers to converge to the convex hull formed by the states of leaders. State formation-containment problems of swarm systems were transformed into asymptotic stability problems, and an explicit expression of the formation reference function was derived. Sufficient conditions for swarm systems to achieve state formation-containment were proposed. Necessary and sufficient conditions for swarm systems to achieve state containment and time-varying state formations were presented respectively as special cases. An approach to determine the gain matrices in the formation-containment protocols was given. Second, output formation-containment control problems for high-order LTI swarm systems were investigated using an static output formation-containment protocol. Output formation-containment problems of swarm systems were transformed into asymptotic stability problems, and an explicit expression of the time-varying output formation reference function was derived. Sufficient conditions for swarm systems to achieve output formation-containment were proposed. An approach to design the output formation-containment protocol was given. It was revealed that state formation-containment, output/state containment, output/state formation control, output/state consensus, and output/state consensus tracking problems can be unified in the framework of output formation-containment problems.

7.2 Future Work

Based on the work of this thesis, there are several interesting and challenging problems on formation and containment control of high-order swarm systems for the further study, which are listed as follows.

First, for the time-varying formation control problems of swarm systems, the collision and obstacle avoidance strategies were not considered in the time-varying formation protocol of this thesis. In practical applications, the capability of collision and obstacle avoidance is essential and may be realized by introducing appropriate attracting and repelling potential functions into the formation protocol. However, the potential functions may result in local minimal solutions and nonlinearities which make the time-varying formation analysis and design problems much more challenging.

Second, only sufficient conditions for high-order swarm systems to achieve formation-containment were derived in this thesis. Whether there exist necessary and sufficient conditions for high-order swarm systems to achieve formation-containment, and how to obtain these criteria are interesting problems need further study.

Third, the topology switching is very common in networked swarm systems. Although consensus control problems and formation control problems for high-order swarm systems with switching topologies have been addressed before, containment control problems and formation-containment control problems for high-order swarm systems with switching topologies, which are complicated and meaningful problems, are still open.

Moreover, theoretical results on time-varying formation control of high-order swarm systems were applied to time-varying formation control of UAV swarm systems, and formation flying experiments were carried out on five quadrotor UAVs in outdoor environment. However, how to apply the theoretical results on containment control and formation-containment control of high-order swarm systems to the UAV swarm systems, and demonstrate the results by experiments are interesting and challenging practical problems.

Finally, in practical application, some of the agents in the swarm systems may encounter certain faults such as faults in communication, sensors, and actuator. The effects of different faults in the agents on the consensus control, formation control, containment control, and formation-containment control of high-order swarm systems have not been investigated extensively.

Index

X. Dong, *Formation and Containment Control for High-order Linear Swarm Systems*, Springer Theses, DOI 10.1007/978-3-662-47836-3

S

U